# Mathematics Manual 1

- Linear Counting -
- Decimal System -
- Memorization of Operations -

Dr. Pamela Zell Lanaro


Montessori Materials Research Foundation
16492 Foothill Boulevard, San Leandro, CA 94578-2105
(510) 278-1115 FAX (510) 278-1577

INTRODUCTION

1. Statement of Philosophy

# TABLE OF CONTENTS

# CHAPTER 1

# INTRODUCTION

# CHAPTER 2

# LINEAR COUNTING

# THE NUMBER RODS

**Materials**

Ten rods differing in length so that the first rod is ten centimeters long and each succeeding rod is ten centimeters longer than the previous rod. Thus, the tenth rod is one hundred centimeters (one meter) in length. The faces remain constant at two centimeters squared. Each rod has segments alternating in colors of red and blue, ten centimeters in each segment.

**Presentation 1**

1. Place the rods on a rug in graded order with the left side of the rods even to each other and against the left edge of the rug.
2. Slide the number one rod down. Say, "This is one." Place the whole hand across the segment and say, "one." Then slide the hand from the left edge along to the right edge and say, "This is one." Have the child repeat the process.
3. Slide the number two rod down. Place the whole hand across the first segment and say, "One." Move to the next segment and say, "Two." Then slide the hand from the left edge to the right edge and say, "This is two." Have the child repeat the process.
4. Continue with the child as long as the child is able to count the segments on a one-to-one correspondence basis. Dramatically and rhythmically counting the segments encourages success.
5. Continue with all ten rods as the child is ready.

**Presentation 2**

1. Place the rods on a rug in graded order with the left side of the rods even to

each other and against the left edge of the rug.

2. Bring the one and two rods down. Repeat Presentation #1.
3. Ask the child to hand you the one. Ask the child to hand you the two. Repeat with creativity. Ask, "What is this?"
4. Continue with three and four. Count out the segments first. Ask the child to locate the three and then the four. Then ask, "What is this?"
5. Continue with all the rods, always reviewing previously learned rods.

**Variation 1**

Place the graded rods at one rug. At a distance, place another rug. Remaining at the rug that is empty, request that the child bring the number rod "one." Then place it on the left edge of the rug to begin the gradation. Continue sequentially through the tenth rod. Have the child count the segments as a check.

**Variation 2**

Place the graded rods at one rug. At a distance, place another rug. Remaining at the rug that is empty, request the that child bring the number rod "five." Then place it on the left edge of the rug to grade the other rods around it. Continue in mixed order through all the rods. Have the child count the segments as a check.

**Variation 3 (Larger - Smaller)**

Place the graded rods at one rug. At a distance, place another rug. Remaining at the rug that is empty, request that the child bring the number rod "five." Then place it on the left edge of the rug to begin the gradation. Say, "Bring me a rod larger than this one." Check the child's accuracy by mapping the five rod to the selected rod. Then count the segments in the selected rod. Make the statement, "is larger than five." Continue with several more requests for larger rods. Then say, "Bring me a rod smaller than the five."

**Variation 4 (One More - One Less)**

Repeat Variation #3, but ask specifically for a rod one larger than the rod before you. Continue asking each time for the rod exactly one larger than the rod indicated. Then ask for one smaller than the rod you indicate.

**Variation 5**

Repeat Variation 4, but ask the child for the rod two less or more than the indicated rod.

**Variation 6**

Repeat Variation #4, but employ the signs for greater than and less than.

**Variation 7 (Addition)**

Place the graded rods on the rug. Move the ten rod down. Just beneath it place the nine rod. Bring the one rod next to the nine rod so that it is as long as the ten rod. Slide your hand along the rod as you say, "Nine plus one equals ten." Continue with eight plus two, seven plus three, six plus four, five plus five (flip the five number rod; if there is confusion bring the red rod out).

**Variation 8 (Addition)**

Repeat Variation #7, but with the nine rod as the base for combinations. Continue with all the rods, using each new rod as the base for combinations.

**Variation 9 (Subtraction)**

Proceed as in Variation #7. When complete, move the ten rod, the nine rod, and the one rod toward the top of the rug. Slide your hand along the length of the appropriate rod while saying, "Ten take away one (remove the one and place at the bottom of the rug) leaves nine." Then move the ten rod below the nine rod and place the eight rod

and the two rod below the ten rod. Continue as with the nine and one rods. For the five, use the red rod and the number rod.

**Variation 10 (Count Backward)**

Grade from the longest to the shortest rod and count backwards.

**Aims**

**Indirect**

Learn the names of the sets 1 - 10.

Learn that each quantity has a separate name, and that each name stands for a separate quantity.

Memorize the sequence of numbers 1 - 10.

Make the correct response between spoken number and its quantity.

Learn the concept of linear quantity.

**Direct**

Develop the concept of number order.

Develop skills in exactness.

Independence.

Coordination of one's body movements.

Problem Solving techniques.

Concentration.

Introduce the concepts of addition and subtraction.

Preparation for metric measurement.

**Control of Error**

Visual.

**Age**

2-1/2 to 5.

**NOTE:**

**Presentation**

Always go back to one before introducing new numbers. The relationship between the one rod and each of the other rods may be revealed by mapping the one rod to the other rods.

**New Math**

Number Rods

Learning the quantity and their symbols from 1 to 10.

Learning the sequence of numbers from 1 to 10.

Introduction to the idea of the Base Ten System.

Introduction to the communtative and associative properties in addition.

Introduction to the concept of a standard of measure.

Introduction to the Pythagorean Theorem.

Red Rods

Relationships

Equality with the first red rod.

Ratios.

Dissimilarities and inequalities.

Greater than, less than.

Equivalents.

Set Concepts

Mapping.

Successor sets.

Empty set.

Number Set

- Introduction to Base Ten Concept.
- Commutative and Associative Properties.
- Regrouping.

# THE SANDPAPER NUMERALS

## Materials

A set of numerals from one through nine cut out of sandpaper and mounted on painted masonite about 8 cm. by 10 cm. Also, a zero that is not presented at this stage (Neinhuis manufactures both a printed and a cursive set of numerals). The cursive set is recommended because the child traces the numeral more easily.

## Presentation

1. Place the numeral cards on a table.
2. Place the box in the upper left hand corner.
3. Remove the first three numerals from the box (one, two, and three). Place the numerals face down in a stack to the left and below the box.
4. Remove the one and position it in front. Hold the card in position with the subdominant hand. With the dominant hand trace the numeral and say its name.
5. Allow the child to repeat the procedure.
6. Continue in this manner to introduce the numerals two and three, each time turning the preceding numeral over and sliding it to the left. (This is a visual control of error.)
7. When all three numerals have been introduced, turn all three numerals over (so that they may face upward.) Begin the second period of the Three-Period Lesson, remembering to lead with the last introduced numeral. Continue to the third period of the Three-Period Lesson if appropriate.
8. Continue in this manner introducing the rest of the numerals with the second and third periods of the Three-Period Lesson.

**Variation 1**

The child may be interested in forming the numerals by tracing in a shallow layer of farina on a tray or with water on a chalkboard. With a tool, such as chalk on a chalkboard or pencil on a paper, the child's skill will be further increased.

**Variation 2**

The sandpaper numerals may be matched to the wooden numeral cards or to a strip that has the numerals duplicated in sequence. A wall chart of the numerals is recommended.

**Variation 3**

The numerals may be placed on one rug and at another distant rug; then the teacher may request that the child bring the numerals, one at a time, as they are requested. The numerals should be in sequence at the first rug.

**Variation 4**

Place the sandpaper numerals in sequential order on a rug. At another rug, place the other set of numerals in mixed order in a stack face down (face down because the children delight in the suspense of it). Turn the first card over and have the child return to the first rug to select the appropriate numeral.

**Aims**

**Indirect**

Learning to count from one to ten.

Recognition of the numerals one to ten.

Development of body coordination.

Making the appropriate association between written symbol and spoken word.

**Direct**

Preparation for writing.

Preparation for the association of quantity and number.

Giving the key to the world of written numerals.

**Age**

2-1/2 to 4.

# THE NUMBER RODS AND NUMERALS

**Materials**

Number rods one to ten. Ten numeral cards with the numerals one through ten printed in red.

**Presentation**

1. Note the placement of the rods in the box with the left edges aligned.
2. Remove the numeral ten from the box and place it before the child and state, "This is ten." Then replace the numeral in the box.
3. Remove the numeral one from the box and say, "One."
4. Then continue in this manner with the numeral two and the two number rod.
5. Because of the prior exercises, the child will be able to continue with the material independently, associating the symbol with the quantity.
6. When all numerals and rods have been associated, the numerals may be placed in order in the box.

**Variation 1**

Place the rods on a rug in mixed array with the left edges aligned. Place the numeral cards in sequential order from left to right at the bottom of the rug. Select a rod and request that the child select the numeral card that goes with the rod. Continue until all rods have the appropriate symbols.

**Variation 2**

Place the rods in sequential order with the left edges aligned. Place the numeral cards in mixed order in a straight line at the bottom of the rug. Select the first numeral card and locate the rod that goes with the numeral. Continue until all numeral cards have

been placed with the appropriate rod.

**Variation 3**

Both number rods and numeral cards are in sequential order. The teacher verbally requests the symbol and quantity, and the child selects both for the association.

**Variation 4**

Place the numeral cards in sequential order on one rug. At another rug, place the number rods in sequential order. Select the first rod and go to the other rug to select the appropriate numeral card to go with the rod selected. Then return with the numeral card to the selected rod.

**Variation 5**

Place the numeral cards in mixed array in a straight line on one rug. At another rug, place the number rods in sequential order. Select the first rod. Go to the other rug and, from the mixed array, select the numeral card that will match.

**Variation 6**

Place the numeral cards in mixed array in a straight line on the rug. At another rug, place the number rods in mixed array with the left edges aligned. Select the first rod and proceed to the other rug to select the appropriate numeral card.

**Variation 7 (Addition)**

Place the numeral at the lower left of the rug in sequential order face down. Place the rods in sequential order. Remove the ten rod and place it below the one rod. Say, "This is ten." Remove the top card from the stack of numeral cards (the ten) and place it to the right of the ten rod, and repeat, "Ten." Then select the nine rod and place it

below the ten rod and say, "Nine." Remove the one rod from the array of rods and place it to the right of the nine rod and just below the ten rod. Say, "One." Remove the numeral one card from the bottom of the stack and say, "One." Gesture and say, "Nine plus one equals ten." Remember to remove the cards and place them face down to the ricrht. Continue making combinations to ten. (The plus sign and equal sign may be introduced in this exercise at a later time.)

**Variation 8 (Addition)**

Continue as above in #7, but with addition combinations of 9, 7, 6, etc.

**Variation 9 (Subtraction)**

When addition formations have been made, then begin to remove rods. Count the quantity that is contained in the two rods combined. Then remove a rod of a specific quantity and determine the quantity that remains. "We have ten; then we remove five and have five remaining." Continue with all the addition formations to remove a rod.

**Aims**

**Direct**

Order, concentration, coordination, independence.

Association of quantity and symbol.

**Indirect**

Preparation for addition and subtraction.

**Age**

3 to 5.

# THE SPINDLE BOXES

### Materials

Two boxes divided into five compartments each. The first box contains the numerals 0 through 4 printed above the compartments. The second box contains the numerals 5 through 9 printed above the compartments. Two baskets, the first basket contains ten spindles, and the second basket contains thirty-five spindles. A container holding 9 rubber bands, ribbons, etc.

### Presentation

1. Point to the zero and say, "This is zero." Then continue pointing to each numeral from left to right, naming each one.
2. Point to the numeral one and say, "One." Then remove one spindle from the basket, place it in front of the one compartment, and say, "One." Grasp the spindle dramatically, saying, "One" as you place it in the one compartment. Then point to the numeral two and say, "Two." Count out one spindle from the basket and place it in front of the two compartment and say, "One." Remove another spindle from the basket and place it in front of the two compartment and say, "Two." Grasp the two spindles and say, "Two." Place a rubber band around them as they are placed in the two compartment. Continue through nine.
3. Point to the zero and say, "Zero." Point to the basket and say, "The basket is empty." Point to the numeral zero and say, "Zero stands for the empty set."
4. The compartments may be emptied by removing the spindles from each compartment individually and placing the spindles in the basket. Or the spindles may be removed from each compartment individually and

placed in front of the compartments while stating the number. They are individually counted back into the basket.

**Variation 1**

The spindles may be replaced by forty-five other objects to lend interest to the material.

**Variation 2**

Many group activities may be centered around zero. Clap three times. Clap two times. Clap zero times.

**Variation 3**

All math materials that are teacher-made should include the zero.

**NOTE:**

In Presentation #1 the child counts the nine spindles for the last compartment and does not simply group the remaining spindles knowing the remaining spindles belong in the last compartment leaving the basket empty.

**Control of Error**

There are only the correct quantity of spindles. The child will have more spindles or less spindles remaining if there has been an error.

**Aims**

**Indirect**

To introduce zero.

The child makes the set of spindles, using the units.

Learn zero as the empty set.

Sequence of numerals is shown to the child.

**Direct**

Introduction to all the symbols used in our numerical system.

Development of order, concentration, coordination, independence and exactness.

**Age**

4 to 5 years.

# THE NUMERALS AND COUNTERS

**Materials**

A set of numeral cards 1 to 10. Or cutout numerals 1 - 10. A box of 55 plastic counters.

**NOTE:**

It is suggested that numeral cards be introduced first to reduce the confusion that the child may have with the cutout numerals and the upside down and backwards possibilities.

**Presentation**

1. Remove the numeral cards and place them in a straight line. Look at the line from left to right to locate the 1. Place the one to the left. Proceed in this way to locate each succeeding numeral, 1 through 10.
2. When the numerals have been sequenced, begin with the 1 and say, "One." Then count the counters, saying "One," and placing it below the numeral one.
3. Point to the two and say, "Two." Count out one counter, say, "one," and place it below the two; then remove another counter and say "two" as you place it below the two and to the right of the previous counter.
4. Continue with all the numerals and counters, taking care to place the counters in sets of two below the numeral.
5. Return the counters to the box. Stack the numeral cards in order.

**Variation 1**

If the child has difficulty in sequencing, the numerals printed on a long strip of cardboard may be used for matching. This isolates the level of difficulty.

**Variation 2**

As an introduction to the use of the cutout numerals, trace the numerals on a long strip cardboard, and match the cutout numerals to the tracings.

**Variation 3**

Proceed as in Presentation #1, using the cutout numerals instead of the numeral cards.

**Variation 4**

When the child has sequenced the numerals and counters, ask the child to indicate the numerals that have complete sets of two counters. Then state that the numerals composed of exact sets of two are called even numerals; the other numerals that do not are called odd numerals.

**Variation 5**

When the numerals and counters are laid out, ask the child to quickly pick up the even (or odd) numerals and name them. Continue creatively with this activity.

**Variation 6**

Gather any quantity of counters, hand them to the child, and request that the counters be laid out in two rows. Is the number odd or even?

**Variation 7**

Have the child place only the even numerals out for making sets of counters.

## Aims

### Direct

Development of order, concentration, coordination, independence and exactness.

To place numerals in their correct sequence.

To verify that the child can make the appropriate grouping for the numerals.

To familiarize (by providing an experience in grouping sets of two) the child with odd and even numerals.

### Indirect

To concretely present multiples and their commutativity (2x3=3x2).

To convey the concept of odd and even numbers.

## Age

4 to 5 years.

# THE MEMORY GAME WITH NUMERALS AND COUNTERS

**Materials**

A basket with slips of folded paper with the numerals 0 -10 written on them. Place a line under the 9 and the 6 so that the child will easily determine the quantity to collect. A basket with 55 objects, such as small cubes.

**Presentation**

1. The children gather around and select a folded slip of paper, look at it, refold it, and place it on the rug before going to the object basket to count out the correct quantity of objects.
2. Upon returning to the rug with the objects, they unfold the paper, and name the numeral, and count out the number of objects.
3. After each child has had a turn and all the slips of paper have been used, the papers may be returned to the basket and the objects to their basket.

**Variation 1**

Proceed as in Presentation and do not bring the object basket. The child may choose objects from the environment to make the set required. The teacher may direct the child to get specific objects.

**Aims**

**Direct**

Development of independence, concentration, coordination, independence and exactness.

Development of memory.

Reinforcement of math material 0 - 10.

**Age**

4 to 5 years.

# LINEAR COUNTING

The child's introduction to numbers begins with 0-10; the materials used are the number rods, sandpaper numerals, spindle box, numerals and counters, and the number memory game.

Linear counting is resumed with:

Short Bead Bars
Teen Board
Ten Board
One Hundred Board and Tiles
One Hundred Chain and Labels
One Thousand Chain and Labels
Squaring Chains and Labels
Cubing Chains and Labels

Typically, linear counting comes after the introduction to the golden bead decimal material, but before the stamp game. [1]

The Teen Board and Ten Board materials follow the Seguin Three-Period Lesson Format.

---

1 Specifically, the Teen Board, Ten Board, Hundred Board, Hundred Chain, and Thousand Chain are presented after the Introduction to the Golden Beads, but before the operations with the golden beads.

# THE SHORT BEAD STAIR

**Materials**

A box of bead bars: one bead colored red, two beads colored green, three beads colored pink, four beads colored yellow, five beads colored light blue, six beads colored lavender, seven beads colored white, eight beads colored brown, nine beads colored dark blue.

**Presentation**

1. Select the one bead bar. Say, "This is one."
2. Select the two bead bar. Count from left to right with finger or notched card. Say, "One, two. This is two."
3. Select the three bead bar. Count with finger or notched card. Say, "One, two, three. This is three."
4. Using the Seguin Three-Period Lesson review this lesson.
5. Continue in this manner to nine according to the pace of the child.

**Aim**

**Direct**

Development of order, concentration, coordination, independence and exactness.

**Indirect**

To prepare the child for the use of the bead bars with the teen and ten board.

**Age**

4 years.

**NOTE:**

The colored bead bars clearly distinguish each number, 1-9.

# SNAKE GAME: SEARCH FOR TEN

**Materials**

A box containing colored bead bars for the numerals one to nine. There should be eight to ten bars for each number.

one - red

two - green

three - pink

four - yellow

five - blue

six - lavender

seven- white

eight - brown

nine - dark blue

A box with a set of black and white bead bars. All the bead bars through five are black. From six through nine the bead bars are black for the first five beads and white for successive beads. This allows for ease in identifying the quantity on the bead bar. A box of ten bead bars. A notched strip of cardboard.

**Presentation 1**

1. Remove the black and white bead bars and place them to the upper right of the work area in graded order.
2. Select an arrangement of colored bead bars and lay them "snakewise" across the work area, taking care to place successive bead bars such that they will add up to an even ten.
3. Using the notched cardboard strip, count from left to right to ten. Because the beads have been so arranged to come evenly to ten, the black and white bead

bars will not be used. Upon counting to ten, place a golden ten-bead-bar above the colored bead bars, say, "ten", remove the colored beads, and place them to the upper right of the work area. Continue in this manner until all of the colored bead bars have been exchanged for golden bead bars. Say, "I have changed the colored snake to a golden snake."

4. Verification. "Let's see if I have counted correctly." Line up the ten bead bars to the left. Line up the colored bead bars to the right, grouping like colors to gether. Next to each ten place a combination of colored bead bars that will add up to ten. When a one-to-one correspondence has been made between the colored bead bars and the ten-bead-bars, the verification is complete.
5. Return the bead bars to their boxes and to the tray.

**Presentation 2**

Proceed as in Presentation #1 through step #5. Then select at random colored bead bars to make the "snake." Using the notched cardboard strip, count to ten. Place a golden ten bead bar above the colored bead bars. Count the remaining beads on the colored bead bar, and place a black and white bead bar to "hold the place" of the colored bead bar that shall be removed. Remove the counted colored bead bars and place them in the upper right work area. Continue to count the colored bead bar snake in this manner, using the black and white bead bars to "hold the place" of colored bead bars that must be removed. Verify as in Presentation #l and exchange colored bead bars as needed from the box of stored colored bead bars.

**NOTE:**

In the event there are colored bead bars remaining in the snake that do not add up to 10, exchange them for a black and white bead bar.

## Aim

### Direct

Development of order, concentration, coordination, independence and exactness.

To explore the possible combinations of ten.

### Indirect

To learn the addition facts.

## Age

4 to 5 years.

# TEEN BOARD: QUANTITY

**Material**

A set of nine ten-bars in a box and one set of colored bead bars, one through nine.

**Presentation**

1. Place the ten bars at the upper left of the work area. Count the quantity of golden beads in a bar to reinforce the quantity.
2. Place the colored bead bars to the right of the ten bars in sequence from one to nine. Beginning with the unit bead-bar, remove it from the sequence and count it. Then replace it in the sequence and select the next bead bar on through to the nine. The counting of the beads may be done with a notched card as in the snake game.
3. Select one ten bar and place it vertically on the rug. Select the unit bead and place it to the right of the ten bar. Say, while pointing to the appropriate bead, "Ten and one more are eleven."

   Select another ten bar and place it vertically beneath the other ten bar. Select the two bead bar and place it to the right of the ten bar. Say, while pointing to the appropriate bead, "Ten and two more are twelve."
4. Then use the two quantities with the Seguin Three-Period Lesson.
5. When the child is successful, remove the one and two bead bars from the ten bars and place them one the rug. Ask the child to make eleven and to make twelve with the bead bars.
6. Continue in this manner through to nineteen according to the pace of the child.

**Aim**

**Direct**

To learn the teen quantities and their sequence.

**Language**

Teen numerals.

**Age**

4 years.

**NOTE:**

1. For independent work, picture cards of bead quantities may be made; the child uses beads to make the quantity depicted.
2. The moveable alphabet may be used to spell the teen numbers.

# TEEN BOARD: SYMBOLS

**Material**

Long rectangular box containing two wooden boards with six wooden strips running horizontally across the boards and the numeral ten painted in each slot. A set of numeral cards, one through nine; each numeral card fits in a slot.

**Presentation 1**

1. Remove the wooden boards and place them vertically, one beneath the other, with the board with the empty slot at the bottom.
2. Place the numeral cards in sequence in a stack at the top of the rug to the right of the boards.
3. Point to the numeral ten at the top of the board and say, "Ten." Select the numeral one, slide it from the right side of the board over the zero, and say, "And one are eleven."
4. Point to the numeral ten directly under the eleven and say, "Ten." Select the numeral two and slide it from the right side of the board over the zero and say, "And two is twelve."
5. Then use the eleven and twelve with the Seguin Three-Period Lesson.
6. When the child is successful, remove the numerals one and two. Ask the child to make eleven; ask the child to make twelve.
7. Continue in this manner through to nineteen according to the pace of the child.

**Presentation 2**

Proceed as in Presentation #1, but place the numeral cards in mixed order for the child to determine the position on the Teen Board where each numeral card belongs.

**Aim**

**Direct**

Learn the teen numerals and their sequence.

**Age**

4 years.

**NOTE:**

1. For independent work, numeral cards with the teen numerals may be made; the child selects a card and makes the numeral on the Teen Board.
2. The moveable alphabet may be used to spell the teen numerals.

# TEEN BOARD: QUANTITY AND SYMBOLS

**Material**

The teen boards and numeral cards, colored bead bars 1-9, and nine ten-bars.

**Presentation 1**

1. Place the wooden boards vertically one beneath the other, the board with the empty slot at the bottom.
2. Place the ten bars at the upper left of the work area. Place the colored bead bars in sequence one through nine to the right of the ten bars.
3. Place the numeral cards in sequence, one through nine, in a stack at the top of the work area on the right.
4. Select a ten bar, place it to the left of the ten symbol at the top of the board, and say, "Ten." Select the one bead and place it to the right of the ten bar and say, "And one are eleven."
5. Point to the numeral ten to the right of the eleven beads and say, "Ten." Select the numeral one, slide it over the zero, and say, "And one are eleven."
6. Proceed in this same manner with twelve.
7. Continue, having the child making the quantity and the symbol according to the child's pace.
8. Using the Seguin Three-Period Lesson, ask for the child to indicate particular teen beads and teen numerals.
9. The numerals may be removed and laid out on the work area. The beads may be placed in their original order. Then a request may be made to create a particular teen quantity.

**Presentation 2**

Mix the numeral cards and invite the child to sequence the numerals on the Teen Board.

**Presentation 3**

Select a numeral at random, place it in the first position on the Teen Board, and make the appropriate bead quantity.

**Presentation 4**

Make a number with the beads and ask the child to make the appropriate numeral on the Teen Board.

**Presentation 5**

Introduce the moveable alphabet for the formation of the teen words.

**Variation 1**

For independent work, picture cards of bead quantities may be made; the child makes the bead quantity the teen numerals that go with it each card.

**Variation 2**

For independent work, cards with the teen numerals may be made; the child makes the numeral on the Teen Board and the bead quantity that go with each card.

**Aim**

**Direct**

Development of order, concentration, coordination, independence and exactness.

### Indirect

To associate the teen numerals with the quantity.

### Age

4 years.

# NUMBER RODS - TEEN PRESENTATION

**Material**

The number rods, numeral cards one through ten, and numeral cards eleven through nineteen.

**Presentation 1**

1. Place the number rods in sequence in the upper left work area.
2. Select the ten rod and place it in the center of the work area. Select the one rod and place it to the right of the ten rod, allowing a small space between the two rods.
3. Say, while pointing appropriately, "Ten plus one equals eleven." Then slide the one rod to meet the ten rod.
4. Remove the one rod and place it in sequence.
5. Select the two rod and place it to the right of the ten rod allowing a small space between the two rods.
6. Say, while pointing appropriately, "Ten plus two equals twelve." Then slide the two rod to meet the ten rod.
7. Continue in this manner through to nineteen according to the pace of the child.

**Presentation 2**

1. As in Presentation #1, lay out the number rods.
2. Place the numeral cards sequentially, one through ten, in a stack to the right of the rods. Place the numeral cards sequentially, eleven through nineteen, in a stack to the right of the numeral cards one through ten.
3. Select the ten rod and place it in the center of the work area. Select the one rod

and place it to the right of the ten rod allowing a small space between the two rods.

4. Say, while pointing appropriately, "Ten plus one equals eleven."
5. Select the numeral ten and place it below the ten rod. Select the numeral one and place it below the one rod. Select the numeral eleven, place it to the right of the numeral cards, and say, while pointing appropriately, "Ten plus one is eleven."
6. Remove the one rod and the one and eleven numeral cards.
7. Continue in this manner through to nineteen according to the pace of the child.

**Aim**

**Direct**

Development of order, concentration, coordination, independence and exactness.

**Indirect**

To reinforce the teen boards.

**Age**

4 years.

# TEN BOARD: QUANTITY

## Material

Nine ten-bars.

## Presentation

1. Place the ten-beads to the left of the work area.
2. Select one ten-bar and say, "This is ten."
3. Select a second ten-bar and place it to the right of the first ten-bar and say, "Ten, twenty."
4. Select a third ten-bar and place it to the right of the two ten-bars and say, "Ten, twenty, thirty."
5. Then with three ten-bars use the Seguin Three-Period Lesson to request the child to make the appropriate quantity.
6. Continue with forty through ninety according to the pace of the child.

## Aim

### Direct

To learn the names of the quantities ten, twenty, thirty, forty, fifty, sixty, seventy, eighty, and ninety.

## Language

Ten numbers.

## Age

4 years.

**NOTE:**

For independent work, picture cards of bead quantities may be made; the child makes the quantity depicted.

# TEN BOARD: SYMBOLS

**Material**

Long rectangular box containing two wooden boards with six wooden strips running horizontally across the boards and the numerals 10, 20, 30, 40, 50, 60, 70, 80, and 90 painted in the slots. The final slot is empty.

**Presentation**

1. Remove the wooden boards and place them vertically one beneath the other, the board with the empty slot at the bottom.
2. Point to the numeral ten and say, "This is ten. Point to the numeral twenty and say, "This is twenty." Point to the numeral thirty and say, "This is thirty."
3. With the Seguin Three-Period Lesson review the lesson.
4. Continue in this manner through to ninety according to the pace of the child.

**Aim**

**Direct**

Development of order, concentration, coordination, independence and exactness.

**Indirect**

To learn the names of the numerals 10-90.

**Language**

Ten numerals.

**Age**

4 years.

**NOTE:**

1. For independent work, numeral cards with the ten numerals may be made; the child makes each numeral on the Ten Board.
2. The moveable alphabet may be used to spell the ten numbers.

# TEN BOARD: QUANTITY AND SYMBOLS

**Material**

The ten board and numeral cards, colored beads 1-9, and nine ten-bars. (An alternate material is golden unit beads instead of colored bead bars.)

**Presentation 1**

1. Place the wooden boards vertically, one beneath the other, the board with the empty slot at the bottom.
2. Place the ten bars at the upper left of the work area.
3. Select one ten-bar and place it to the left of the numeral ten at the top of the wooden board. Point to the bead bar and say, "Ten." Point to the numeral ten to the bar's right and say, "Ten."
4. Move the ten-bar down to the left of the numeral twenty. Select another ten-bar and place it next to the first ten-bar. Point to the ten bar and say, "Ten." Point to the second ten bar and say, "Twenty." Point to the numeral twenty and say, "Twenty."
5. Move the two ten-bars down to the left of the numeral thirty. Select another ten-bar and place it next to the two ten-bars. Point to the first ten-bar and say, "Ten." Point to the second ten-bar and say, "Twenty." Point to the third ten-bar and say, "Thirty." Point to the numeral thirty and say, "Thirty."
6. Continue, having the child make the quantity next to the appropriate numeral according to the child's pace.
7. Using the Seguin Three-Period Lesson, ask for the child to make various tens in random order.

**Presentation 2 (Using Golden Bead Units)**

1. Place the unit-beads to the right of the ten-bars. Place the numeral cards in sequence, from one through nine, to the right of the wooden boards.
2. Select a ten-bar and place it to the left of the numeral ten. Select a unit-bead and place it to the right of the ten bar and say, "Eleven."
3. Select the "one"numeral card, place it in the slot over the zero of the ten, and say, "Eleven."
4. Remove the numeral one card and place it face down to the right of the stack of numeral cards.
5. Continue in this manner through to nineteen.
6. After nineteen, place the unit-bead below the nine unit-beads that make the nineteen and say, "Nineteen and one more are twenty." Select a ten-bar and place it next to the ten unit-beads so that the relationship may be seen. Then remove the unit-beads so that the relationship may be seen. Then remove rthe unit-beads and move the two ten-bars down next to the numeral twenty.
7. Continue with the twenties, going sequentially from twenty-one to twenty-nine. At twenty-nine, place a unit-bead below the nine units and demonstrate the relationship between ten units and a ten bar. Then exchange for a ten bar. Continue through to ninety-nine.
8. Upon arriving at ninety-nine, add one more unit-bead, exchange for a ten-bar. Then exchange all ten ten-bars for a hundred square and place it in the blank section of the Ten Board.

**Aim**

**Direct**

Development of order, concentration, coordination, independence and exactness.

**Indirect**

To associate the ten numerals with the quantity.

**Age**

4 years.

**NOTE:**

1. It would be a rare child, indeed, that would have the span of concentration to continue through to ninety-nine in one sitting. However, it is conceptually advantageous to show the child the exchange from the nine to the next ten a couple of times (19, 29) at one lesson.

   To spark interest, some children enjoy recording each numeral as they make it on a roll of paper.
2. For independent work, picture cards of bead quantities may be made; the child makes the bead quantity and the numeral on the Ten Board that go with each card.
3. For independent work, cards with the numerals may be made; the child makes the numeral on the Ten Board and the bead quantity that go with each card.

# ONE HUNDRED CHAIN AND LABELS

**Material**

A chain of ten, ten-bars connected by links. A hundred square. A set of arrows with numerals:

1-9 green arrows (small)

10-90 blue arrows (medium)

100 red arrow (large)

**Presentation**

1. Carry the chain with great care, one hand at each end of the chain.
2. Stretch the chain out vertically.
3. Place the hundred square above the chain.
4. With care and dexterity, move the ten-bars up, one at a time, to form a hundred square.
5. Superimpose the hundred square over the hundred square formed by the hundred chain.
6. Replace the hundred square.
7. Point to the hundred square and say, "This is one hundred." Point to the hundred square formed by the hundred chain and say, "This is one hundred."
8. Stretch the hundred chain out again.
9. Place the arrows in order: 1-9, 10-90, 100, to the right of the chain.
10. Begin to count the beads, beginning at the top and placing the appropriate arrow by the bead at the end of each set.
11. The arrows are green through nine.
12. At ten, a blue ten arrow is placed next to the bead.
13. The counting may proceed by ones, counting each bead with the small, green

arrows. Or the counting may proceed by tens, using the blue arrows.

10, 10-20, 10-20-30, 10-20-30-40, 10-20-30-40-50, 10-20-30-40-50-60, 10-20-30-40-50-60-70, 10-20-30-40-50-60-70-80, 10-20-30-40-50-60-70-80-90.

14. Upon counting by ones or by tens, place a red hundred arrow next to the hundredth bead.
15. Place the hundred bead square at the bottom of the chain.
16. Walking along the chain, count in sequence from ten through one hundred.
17. Walk back counting backwards, one hundred through ten.
18. Name a quantity, such as sixty-four, and ask the child to point to the linear quantity representing sixty-four.
19. Point to a particular bead and ask the child to name the quantity.

**Aim**

**Direct**

To learn the numbers through one hundred, by ones and by tens.

**Age**

5 years.

**NOTE:**

1. The child has experienced one hundred beads in a square, but this is the first experience with the quantity stretched out so that each ten bar is easily identified.
2. Labels may be made (e.g., 1 x 10 = ) to place by the appropriate set of ten bars.

# ONE THOUSAND CHAIN AND LABELS

**Material**

A chain of one hundred ten-bars connected by links, ten hundred-squares, and a thousand-cube. A set of arrows with numerals:

| | |
|---|---|
| 1-9 | green arrows (small) |
| 10-90 | blue arrows (large) |
| 100-900 | red arrows (larger) |
| 1000 | green arrow (largest) |

**Presentation**

1. Carry the chain with great care.
2. Stretch the chain as much as possible to dramatically illustrate the great quantity in one thousand beads.
3. Place the ten ten-bead squares above the chain.
4. Place the thousand bead cube to the right of the ten bead squares.
5. With great care and dexterity, move the ten-bars up, one at a time to form a ten bead square.
6. Superimpose a ten bead square over the ten bar square. Then place the ten bead square to the right of the square.
7. Continue in this manner through all the ten bars, forming ten squares and superimposing a bead square, then place them to the right. In this way two long rectangles are formed: one of ten bead bars, the other of bead squares.
8. "How many squares do we have?"
9. Beginning with the last bead square, stack the squares while counting. When you arrive at "ten squares" stacked next to the thousand bead cube, point to the thousand cube and say, "Ten hundred equals one thousand."

10. Then with care and dexterity, stretch the thousand chain out.
11. Place the arrows in order: 1-9. Count the beads, beginning at the top and placing the appropriate arrow at the bead. The arrows are green through nine.
12. Place the arrows in order: 10-100.
13. Count the beads by ones or by tens, placing the appropriate blue arrows to the right of the bead bars. When one hundred is reached, place the red arrow by the ten bead bar and place the ten bead square to its right.
14. Place the arrows in order: 110-200.
15. Count the beads by tens, placing the appropriate red arrow to the right of the bead bars. When two hundred is reached, place the red arrow by the ten bead bar and place the bead square to its right.
16. Continue in this manner to one thousand. At one thousand, also place the thousand bead cube to the right of the ten bead square.
17. Walking along the chain, count in sequence from ten through one thousand.
18. Walk back counting backwards one thousand through ten.
19. Name a quantity, such as six hundred fifty-eight. Ask the child to point to the linear quantity representing six hundred fifty-eight.
20. Point to a particular bead and ask the child to name the quantity.

**Variation**

Make up tags with numbers, such as 234. The child places the tag at the particular bead.

**Aim**

**Direct**

Development of order, concentration, coordination, independence and exactness.

**Direct**

To learn the numbers through one thousand, by ones and by tens.

**Age**

5 to 6 years.

**NOTE:**

1. It is important to allow the child to stretch the chain to its full length so the child will have a sense of linear quantity.
2. The long chain is replaced on the frame by selecting alternate large rings in the chain. These alternate rings are placed on the bead cabinet hooks.
3. Make a long mat of felt on which to lay chains.

# ONE HUNDRED CHAIN AND ONE THOUSAND CHAIN

**Material**

One hundred chain and labels; one thousand chain and labels.

**Presentation**

1. Proceed as in the Presentation for the one hundred chain. Do not use the ten bead squares.
2. Proceed as in the Presentation for the one thousand chain, placing the chain parallel to the hundred bead chain. Do not use the ten bead squares.
3. Upon completion, compare the two chains.

**Aim**

**Direct**

Visualize the difference in the quantity one hundred and one thousand.

**Age**

5 years.

# ONE HUNDRED BOARD AND NUMERALS

### Material

A board with a wood frame with a ten by ten matrix on the top. The numeral one is printed in the upper left corner. Small squares with numerals printed one through one hundred. Initially, the numeral squares are organized into sets of ten: 1-10, 11-20, 21-30, 31-40, 41-50, 51-60, 61-70, 71-80, 81-90, 91-100. One through one hundred control chart.

### Presentation 1

1. Take the hundred board to the work area.
2. Select the first set of numeral squares, 1-10.
3. Lay the numeral tiles in mixed array in a horizontal line below the hundred board.
4. Choosing among the numeral squares, from left to right, select numeral one.
5. Place the numeral one in the upper left corner of the hundred board.
6. Choosing among the numeral squares, from left to right, select numeral two.
7. Place the numeral two in the upper left corner to the right of the numeral one.
8. Continue in this manner placing the numerals in sequence from one through ten.
9. Select the second set of numeral squares, 11-20.
10. Continue as with the first set of numerals through all sets.

### Presentation 2

1. Continue as in Presentation #1.
2. Upon laying the first set of numerals squares in mixed array in a horizontal

line, select the first numeral.

3. Begin to count from the first position on the hundred board, and continue until the correct position has been reached for the numeral selected. Place it in its position on the board.
4. Continue to select a numeral square from the mixed array. Determine its cor rect position by counting across the positions.
5. Continue in this way with each set.

**Presentation 3**

Continue as in Presentation #1. However, all sets are laid out. Randomly select a numeral square. Determine its position on the matrix by counting the tens down and then the units over.

**Aim**

**Direct**

To reinforce the sequence of numerals one through one hundred.

**Age**

4 to 5 years.

# SQUARING CHAINS: ONE THROUGH TEN

**Material**

Wooden frame with ten hooks to hold the following bead chains:

1. One red bead bar, red label 1
2. Two green bead bars, green labels 2, 4
3. Three orange bead bars, orange labels 3, 6, 9
4. Four yellow bead bars, yellow labels 4, 8, 12, 16
5. Five light blue bead bars, light blue labels 5, 10, 15, 20, 25
6. Six lavender bead bars, lavender labels, 6, 12, 18, 24, 30, 36
7. Seven white bead bars, white labels 7, 14, 21, 28, 35, 42, 49
8. Eight brown bead bars, brown labels 8, 16, 24, 32, 40, 48, 56, 64
9. Nine dark blue bead bars, dark blue labels 9, 18, 27, 36, 45, 54, 63, 72, 81

Squares for each of the chains

**Presentation 1**

1. Select a chain. Usually four, five, or six is best for illustration of the procedure.
2. Carry the chain with great care, one hand at each end of the chain.
3. Place the chain stretched out vertically.
4. Place the square at the top of the chain.
5. With care and dexterity, move one bar up at a time to form the square.
6. Superimpose the bead square over the square formed by the chain.
7. Replace the square at the top of the squared chain.
8. Stretch the chain out again.
9. Place the arrows in order to the right of the bead chain.
10. Count the beads, beginning at the top and placing the appropriate arrow by

the bead at the end of each set. Place arrows 1-6, then 12, 18, etc., counting by 1.

11. After placing the final arrow, which is the square of the bead chain, place the bead square to the right of the arrow indicating the square.
12. Walking along the chain, count in sequence from first set to the last set of bead bars.
13. Walk back counting backwards.

**Presentation 2**

1. Proceed as in Presentation #1, with the chain and labels.
2. Select another chain and place it parallel to the first chain. Proceed as in Presentation #2, with the chain and labels.
3. Ask the child to find a particular numeral in all the chains. (Eg.: 12).

**NOTE:**

This presentation allows the child to make a visual comparison between squaring chains. The child may note which numerals recur in the various chains. (12 occurs in the 2, 3, 4, and 6 chains.)

**Aim**

**Direct**

Reinforce linear counting.

Begin skip counting.

Experience the relationship between the quantity differences of the squares.

**Age**

4 to 5 years.

# CUBING CHAINS: ONE THROUGH TEN

**Material**

Wooden bead cabinet with top ledge to hold the cubes from left to right: ten cube, nine cube . . . one cube. Hooks extending down from the top ledge of the cabinet to hold the cubing chains from left to right: ten cubing chain, nine cubing chain . . . one cubing chain.

From the right side of the cabinet, ledges extend out in increasing lengths so that the tenth ledge forms the bottom border of the bead cabinet. The top ledge on the right contains the one square and one squaring chain. And each ledge contains the next number in sequence down to the bottom ledge, which contains the ten squares and the ten squaring chain.

Labels organized by cubing chains.

**Presentation**

1. Select a chain. Usually four, five, or six is best for illustration of the procedure.
2. Remove the chain from the cabinet by removing each loop from the top hook and placing it in the dominant hand. Grasp the other end in the subdominant hand if bead quantity permits, or drape the beads across the subdominant arm.
3. Carry the chain with great care.
4. Stretch the chain stretched out vertically.
5. Remove the squares of the chain, one at a time, from the bead cabinet and place them in a stack at the top of the chain.
6. Remove the cube of the chain and place it to the right of the stacked squares.

7. With great care and dexterity, move one bar of the chain up at a time to form a square.
8. Superimpose the bead square over the square formed by the chain.
9. Place the square to the right of the square formed by the bead bars.
10. Continue in this manner through all the bars of the chain, forming a square and superimposing a bead square, then placing the bead square to the right. In this way, two long rectangles are formed: one of bead bars and one of bead squares.
11. "How many bead squares do we have?"
12. Beginning with the last bead square, stack the squares while counting. When the cube is formed with the squares say, "(x) squares equal a cube."
13. With care and dexterity, stretch the cubing chain out.
14. Place the arrows in order.
15. Count the beads, beginning at the top and placing the appropriate arrow at the bead. Continue until the end of the chain.
16. Walking along the chain, read the labels in sequence.
17. Walk back and read the labels backwards.

**Aim**

**Direct**

Development of order, concentration, coordination, independence and exactness.

**Indirect**

To learn the numbers in linear counting as they are arranged in subsets.

To memorize the multiplication facts through skip counting.

**Age**

5 to 6 years.

# CUBING CHAINS LAYOUT

**Material**

All the cubing chains and labels.

**Presentation**

1. Select the one cubing chain. Lay it vertically and place the appropriate label to its right.
2. Select the two cubing chain. Lay it vertically and parallel to the one cubing chain. Place the appropriate labels.
3. Continue with all the cubing chains.
4. Upon completion, note the difference in the lengths of the cubing chains.
5. Note the recurrence of certain numerals along the chains. (Ex. 12.)

**Aim**

**Direct**

Note the difference between the quantities in a cube.

**Indirect**

Factoring from a product.

**Age**

6 years.

# CHAPTER III

# THE DECIMAL SYSTEM

# INTRODUCTION TO THE DECIMAL SYSTEM: QUANTITY

**Materials**

A tray on which is placed, from right to left: one golden bean in a caster, one ten-bar of golden beads, one hundred-square of golden beads, and one thousand-cube of golden beads.

**Presentation 1**

1. Place the golden bead tray on a rug or at a table.
2. Remove the unit bead from the tray and say to the child, "This is one unit." Encourage the child to handle the unit bead to experience its singularity.
3. Place the unit bead to the side, and remove the ten bead-bar from the tray. Say, "This is one ten." Allow the child to count the ten beadbar. Proceed with the Seguin Three-Period Lesson.
4. Continue with the hundred square and thousand cube as indicated above.
5. When the child has experienced success with this activity, the wooden hun dred and wooden thousand may be introduced as representing the hundred golden beads and the thousand golden beads.
6. Replace the beads on the tray in order from right to left, units to thousands.

**Presentation 2**

On a tray, have ten units, ten tens, ten hundreds, and one thousand. Lay out each unit, counting as they are laid in a row. When ten is reached of the unit beads, see if the child sees the relationship between the ten units and the one ten. Point out the relationship if the child does not see it. Continue with counting out ten of the tens and its relationship to the hundred. Then count out ten hundreds and observe its relationship to the thousand.

**Presentation 3 (Decimal Layout in Beads)**

Have forty-five units, tens, hundreds, thousands. Beginning with units, place one in the upper right corner of the work area; directly below it place two units, and continue to nine. To the left of the one unit, place one ten and continue with two tens below the one ten, counting through nine tens. Continue with hundreds and thousands.

**Presentation 4**

Have a tray with a quantity of golden bead units tens, hundreds, and thousands.

1. Ask the child to gather 3 units. Count with the child for accuracy. Select several more from the units for the child to gather.
2. Continue to the tens, hundreds, and thousands in the same manner.
3. Continue with the same request, but more randomly from one category to another, i.e., hundreds, units, thousands, tens.
4. Request two categories at the same time, "Bring me two tens and five units." Group categories: units-tens, tens-hundreds, hundreds-thousands.
5. Request two categories at the same time, but do not group categories, i.e., ones-hundreds.
6. Continue with three categories, and then four categories.

**Aim**

**Direct**

Learn the names of the quantities unit, ten, hundred, and thousand.

Introduction to the decimal system.

**Indirect**

Preparation for the written symbols in the decimal system.

**Age**

4 years.

# INTRODUCTION TO THE DECIMAL SYSTEM: SYMBOLS

**Materials**

A tray on which is placed, from right to left, a card with the numerals for one unit, a card with the numerals for one ten, a card with the numerals for one hundred, and a card with the numerals for one thousand.

**Presentation 1**

1. Place the decimal tray on a rug or on a table.
2. Remove the numeral one card and say, "This is the way we write one unit." Remove the numeral ten card and say, "This is the way we write one ten." Proceed with the Seguin Three-Period Lesson.
3. Continue with the hundred card and the thousand card, as indicated above.
4. Ask the child how many zeros are in the ten, the hundred, the thousand.
5. Replace the numeral cards on the tray in order from right to left, ones to thousands.
6. Return the material to the shelf in the manner indicated in step #3.

**Presentation 2**

The complete decimal symbols are needed. Count systematically through the numeral cards, beginning with units. When nine is counted, say, "What comes after nine?" Then begin the tens category. At nine tens, ask the the question, "What comes after nine tens?" Then begin the hundred category to nine hundreds and ask, "What comes after nine hundreds?"

**Presentation 3 ("Asking" or "Fetching")**

Lay out the full set of decimal numeral cards with the unit cards to the right and the tens to the left of the units, etc.

1. Ask the child for "five units." Ask for several symbols from the units.
2. Continue in this manner through the tens, hundreds, thousands.
3. Request two categories at the same time, but do not group categories, i.e., ones-hundreds.
4. Continue with three category, and then four categories.

**Aims**

**Direct**

Learn the symbols for the quantities of units, tens, hundreds, and thousands.

**Indirect**

Preparation for the association of system number and numerals in the decimal system.

**Age**

4 years.

# THE DECIMAL SYSTEM: ASSOCIATION OF QUANTITY AND SYMBOL

**Materials**

A tray with the numeral cards and golden beads for one unit, one ten, one hundred, and one thousand.

**Presentation 1**

1. Place the one unit bead off the tray and say, "This is one unit." Place the one-unit card above the unit bead and repeat, "This is one unit." Continue with the ten bead-bar and the ten numeral card, then proceed with a three-period lesson. Continue in this manner with the hundreds and thousands.
2. Return the materials to the tray in proper order with the units to the right.

**Presentation 2 ("Bird's Eye View")**

Lay out a complete set of golden beads and the numeral cards, beginning with one unit-bead in the upper right corner of the rug and the one unit numeral card placed to the bead's left. Point to the one unit and say, "This is one unit." Point to the one unit numeral card and say, "This is one unit." Proceed with the units, placing the unit beads and numerals in a row. Then continue in the same manner with the tens, hundreds, and thousands.

**Presentation 3 ("Crises of Ten")**

Lay out a complete set of golden beads and the numeral cards, beginning with one unit bead in the upper right corner of the rug and the one unit numeral card placed to the bead's left. Point to the one unit and say, "This is one unit." Bring the one unit down, add one to it, and say, "This is two units." Continue bringing down the unit

beads to the numeral nine. Then add one more bead (which makes ten) and exchange the ten units for one ten bar; place the ten bar at the top next to the ten numeral card.

**Presentation 4**

Lay out the complete set of decimal numeral cards on a rug. Place a quantity of beads from each category on the child's tray in proper order. Request that the child count the beads and select the decimal numeral card which is appropriate to the bead quantity. When the child has completed the work, check its accuracy by having the child count out each category and place the decimal numeral card above it.

**NOTE:**

If a category of beads is not placed on the child's tray, no decimal numeral cards would be placed on the tray for that category. This "zero" situation will be emphasized when the child counts the materials out to the teacher.

**Presentation 5**

Lay out the complete set of decimal numeral cards on a rug. Select numeral cards for the child to place on a tray, then have the child get the number of beads indicated (2000, 400, 50, 30).

**NOTE:**

1. As a control of error, do not select the same beginning digit for two categories, like 2000 and 200.
2. Check the position of the child's tray for the consistent placement of units to the child's right.
3. The child must replace the beads in the proper category and the numeral cards

where they belong in the decimal card layout.

**Request**

1. Two adjacent categories (30, 6 or 100, 40).
2. Three adjacent categories (600, 40, 7 or 2000, 500, 10).
3. Omit a category (2000, 500, 7).

As a later activity, verbally request that the child bring certain numeral cards, and then the corresponding beads.

**Aims**

**Direct**

Associate the quantity and the symbols for the decimal system.

**Indirect**

Only nine elements exist in any one category.

**Age**

4 years.

# FORMATION OF NUMERAL CARDS

**Material**

A complete set of decimal numeral cards.

**Presentation 1**

1. Place the material on a rug, and lay the decimal numeral cards out in categories from right to left, units to thousands, and from top to bottom, ones to nines.
2. Ask the child to bring a card from the tens and a card from the units (50, 7).
3. Place the unit card over the significant cipher of the ten card. Then "shift" or "tap" the cards so that the unit card falls down to its place over the zero place holder of the ten card. Read the name of the numeral.
4. Continue with other cards from two adjacent categories, then from three adjacent categories, "tapping" the cards into the respective positions.
5. Omit a category and "tap" the cards into place. This dramatizes zero as a place holder.
6. Replace the numeral cards correctly back in the decimal numeral card layout.
7. Replace the numeral cards on the tray or box.
8. Return the material to the shelf.

**NOTE:**

Place the smaller category card under the larger category card. This will reinforce the placement and order of the cards.

## Aims

### Direct

Development of order, concentration, coordination, independence and exactness.

Reinforce the categories of the numerals.

Give the child the "adult" wording for numbers.

### Indirect

It is the place a numeral occupies in the complex of numerals that assigns value.

In a complex of numerals, the zero simply indicates the lack of quantity in that particular category.

## Age

4 to 5 years.

# ADDITION WITH THE GOLDEN BEADS

**Material**

A complete set of golden beads on a shelf arranged from right to left, units, tens, hundreds, and thousands. One complete set of large decimal numeral cards and two complete sets of smaller decimal numeral cards. One large rug or three smaller rugs to accommodate the numeral cards. Three trays with a coaster in each to contain the unit beads.[2]

**Presentation 1**

1. Lay out the small decimal numeral cards as in the decimal layout indicated in earlier presentations. To the right of the small decimal cards, place the large decimal cards in the decimal layout order. To the right of the large decimal cards, make space or lay out another rug on which to perform the operation of addition.
2. With two children, each having a tray, proceed individually to the small numeral cards. Ask the child to select numeral cards that are dictated by the teacher, such as 2436. Then proceed to the shelf with the golden beads and have the child select the appropriate quantity of golden beads. Return to the "work" rug and place the numeral cards to the left, saying the numeral as it is placed on the rug. To the right of the numeral cards place the golden beads, repeating the quantity placed on the rug.
3. Repeat this procedure with the second child for the second addend (1243). Place the numeral cards below the first set of numeral cards on the "work" rug, and the golden beads below the first set of golden beads.
4. Then say, "Let's add all our beads together."

---

2 It is important that the child always take a tray to carry quantity.

5. Then move both sets of unit beads down, then both sets of ten beads down, then both sets of hundred beads down, then both sets of thousand beads down.
6. Begin with counting the unit beads. Ask the child to get the large unit card which corresponds to the unit beads counted. Place it below the numeral cards. Continue in this manner with tens, hundreds, and thousands. When completed, read the problem, "Two thousand, four hundred thirty-six plus one thousands two hundred forty-three equals three thousand six hundred seventy-nine."
7. Replace the numeral cards in their respective boxes.
8. Return the material to the shelf in the manner indicated in step #3.

**Presentation 2 (To Strike the Elementary Age Child's Imagination)**

1. Lay out the small decimal numeral cards as in the decimal lay out indicated in earlier presentations. To the right of the small decimal cards, place the large decimal cards in the decimal layout order.
2. On each tray, place numeral cards for the child to match with golden beads.
3. When the child returns with the tray, check the golden beads for accuracy.
4. Place the numeral cards to the right of the decimal layouts, one set beneath the other. Place the golden beads on a cloth together to dramatize addition - putting together.
5. When both trays have been returned and the beads have been placed together in a cloth, tell the child(ren) that addition means to add like things together.
6. Then count out the unit beads and select the large unit card to place below the small unit cards.
7. Proceed in counting and selecting the large decimal card for the tens, hundreds, and thousands.
8. Verbally repeat the process for the child, i.e., "We took 2436 beads and 3241

beads, added them together, and got 5677 beads altogether."

9. Introduce the "plus" sign (+) and the straight line which separates the addends from the sum.
10. Replace the decimal cards in their respective boxes; place the golden beads back on the shelf in their respective categories.
11. Return the material to the shelf.

**Variation 1 (See "Regrouping with Golden Beads")**

Proceed as in Presentation #1, but add numerals in which the unit category will exceed ten: 1436 + 4327. When all the beads have been combined, begin counting the unit beads until arriving at "ten units."

Ask, "What do ten units make?" The sought response is "One ten." Have the child "regroup" or "exchange" the ten units for one ten. Then continue to count the units and get the large numeral card appropriate to the number of unit beads. Continue as in Presentation #l to count the tens, hundreds, and thousands. "Regrouping" or "exchanging" is introduced in this order:

1. Unit category
2. Ten category
3. Hundred category
4. Unit and ten
5. Unit and hundred
6. Ten and hundred
7. Unit, ten, and hundred

Zero in the sum is introduced in this order:

1. Zero in one category
2. Zero in two categories
3. Zero in three categories

Then problems are introduced in which there are zeros in the addends, i.e., 2450 + 2342 = . At the end of this sequence, the child should be presented with a random selection of problems.

**Variation 2**

Proceed as above, but have three addends.

**NOTE:**

When counting the beads, always name the category, i.e., "one unit," "two units," "three units," etc., or "one ten," "two tens," "three tens."

**Aim**

**Direct**

Development of order, concentration, coordination, independence and exactness.

Putting together of quantities to produce a larger quantity.

"Exchanging" or "regrouping" numbers.

**Language**

2450 Addend

2342 Addend

4792 Sum

**Age**

4-1/2 years.

# REGROUPING WITH THE GOLDEN BEADS FOR ADDITION AND MULTIPLICATION

## Material

Quantities of units, tens and hundreds of golden beads, a tray.

## Presentation

1. Select a quantity of bead in each category in excess of ten. Say to the child, "Because of the way we group quantities, we can never have more than nine of one kind."
2. The units are counted to ten, exchanged or regrouped for a ten-bar, and placed with the tens. Continue counting the units until there is a quantity of nine or less.
3. Continue in this manner with the tens and hundreds. Regroup to the higher category each time ten is reached.

## Aim

### Direct

Preparation for dynamic addition and multiplication.

## Age

4-1/2 to 5 years.

**NOTE REGARDING SEQUENCE:**

Multiplication is the logical sequel to addition with the restriction that the addends are the same. However, the usual public school sequence is to follow addition with subtraction, so this sequence is followed here.

# SUBTRACTION WITH THE GOLDEN BEADS

### Material

Golden beads, three trays, two small sets of decimal numeral cards, one large set of decimal numeral cards.

### Presentation

1. Lay out the sets of large numeral cards on a rug and the two sets of small numeral cards to the right.
2. Select a numeral from the large numeral cards and place it on a tray (7648). Select the appropriate quantity of beads to go with the numeral card.
3. Tell a story to the child, such as, "I am the rich mom my with all the goodies. I will share my goodies with the poor children." Place the large (minuend) numeral cards on a rug with the beads to the right.
4. Indicate, with the small numeral cards on a tray, the quantity which the child may select from the large quantity (2315).
5. The 2315 (subtrahend) is placed below the 7648, and the golden beads specified by the smaller numeral cards are brought down from the larger quantity of golden beads.
6. The golden beads remaining at the top are brought below the golden beads of the subtrahend beginning with units. Upon placing the units to the bottom, say the quantity; then get the small unit card and place it below the unit card of the subtrahend.
7. Continue with the tens in the minuend; place them below the tens of the sub trahend and count them, getting the appropriate small numeral card.
8. When this process has been completed say, "We had ________ (minuend), we

took away ________ (subtrahend), and now we have ________ (remainder) remaining."

### Variation 1

Continue as in Presentation #1, but have two subtrahends.

### Verification

Add the subtrahend and the remainder. The sum should equal the minuend.

### Aim

#### Direct

To familiarize the child with the process of subtraction.

### Language

| | |
|---|---|
| 7648 | Minuend |
| 2315 | Subtrahend |
| 5333 | Remainder |

### Age

5 years.

# REGROUPING WITH THE GOLDEN BEADS FOR SUBTRACTION AND DIVISION

**Presentation 1**

1. Arrange the numeral cards as in "Subtraction With The Golden Beads."
2. Select from the large numeral cards a minuend (4364). Place the cards on a tray and place the quantity of beads that go with it on the tray.
3. To the right of the numeral cards, place a work rug with the minuend numeral cards and the beads to the right of the cards.
4. Select from the small numeral cards a subtrahend in which the unit category will be more than the minuend (2136). Place these numeral cards below the minuend cards.
5. From the minuend beads, take away the quantity indicated by the subtrahend cards. Count: "One unit, t-vo units, three units, four units. We need six units, but there are only four units here. How can I get more units? I can regroup a ten for units."
6. Proceed to go to the golden bead cabinet and regroup the one ten for ten units. Return the units to the ones category of the minuend. Continue to count out, "Five units, six units."
7. Proceed with the tens, hundreds, and thousands.
8. To determine the remainder, begin with the units remaining in the minuend. Count the quantity and place it below the subtrahend. Get the numeral card that corresponds to the quantity and place the numeral card below the subtra hend numeral cards.
9. Continue with the tens, hundreds, and thousands.

10. Read the numeral cards upon completion, and restate the problem. "We had 4364 beads, we took away 2136 beads, and now we have 2228 beads remaining."

**Presentation 2**

1. Set up the large and small numeral cards as indicated in the earlier subtraction activities.
2. With the large numeral cards, select the 1000 numeral card and the thousand cube. Place cards and beads to the left of the layout.
3. From the small numeral cards, make the numeral 999. Place it below the 1000 minuend.
4. Begin with the unit category and say, "I wish to take away nine units. There are none available in the minuend." Then go the tens category of the minuend. "There are no tens." Then go to the hundreds category of the minuend. "There are no hundreds." Then go to the thousands. "There is one thousand. How many hundreds in this thousand?" Regroup the thousand cube for ten hundreds. Take one of the hundreds and say, "How many tens are in this hundred?" Regroup the hundred for ten tens. Take one of the tens and say, "How many units are in this ten?" Regroup the ten for ten units.
5. Proceed in the usual manner to subtract. The remainder is one.

With the process of dynamic subtraction introduced to the child, the sequence for introducing problems is:

1. Exchange or regroup in units.
2. Exchange or regroup in tens.
3. Exchange or regroup in hundreds.
4. Exchange or regroup to produce zero in the tens.
5. Exchange or regroup to produce zero in the hundreds.

6. Zero is in the units of the minuend.
7. Zero is in the tens of the minuend.
8. Zero is in the hundreds of the minuend.
9. Zero is in units and tens of the minuend.
10. Zero is in the tens and hundreds of the minuend.
11. Zero is in the units, tens, and hundreds of the minuend.

At the end of this sequence, the child should be presented with a random selection of problems.

**Age**

6 years.

# MULTIPLICATION WITH THE GOLDEN BEADS

**Material**

A complete set of golden beads on a shelf arranged from right to left, units, tens, hundreds, and thousands. One complete set of large decimal numeral cards and three complete sets of smaller decimal numeral cards. One large rug or four smaller rugs to accommodate the numeral cards. Three trays with a coaster in each to contain the unit beads.

**Presentation 1**

1. Lay out the three sets of small numeral cards as in the decimal layout indicated in earlier presentations. To the right of the small decimal cards, place the large decimal cards in the decimal layout order. To the right of the large decimal cards, make space or lay out another rug on which to perform the operation of addition.
2. With three children, each having a tray, proceed individually to the small numeral cards. Ask the first child to select numeral cards that are dictated by the teacher, such as 2312. Then proceed to the shelf with the golden beads and have the child select the appropriate quantity of golden beads. Return to the "work" rug and place the numeral cards to the left, saying the numeral as it is placed on the rug. To the right of the numeral cards, place the golden beads, repeating the quantity placed on the rug.
3. Repeat this procedure with the other two children dictating, the same multiplicand, 2312.
4. Then say, "Each of you has brought the same amount' Here is 2312 one time and 2312 two times and 2312 three times. Let's see how much we have altogether."

5. Then move all the unit beads down, all the ten beads down, all the hundred beads down, and all the thousand beads down.
6. "Let's see how many we have."
7. Begin with counting the unit beads. Ask the child to get the large unit card which corresponds to the unit beads counted. Place it below the numeral cards. Continue in this manner with tens, hundreds, and thousands. When completed, read the problem, "Two thousand, three hundred twelve, one time, two times, three times equals six thousand, nine hundred thirty-six."
8. Explain to the children that when the addends are all the same, then it is a special kind of addition called multiplication.
9. When this process has been completed, say, "We took 2312 (multiplicand) beads three times (multiplier) and altogether we have 6936 (product)."
10. Replace the numeral cards in their respective boxes.
11. Return the material to the shelf.

**Presentation 2**

1. Proceed as in Presentation #1. When complete, introduce the "multiplication" sign (x) and the straight line which separates the multiplicands from the product.
2. Then say, "Since the addends are all the same, there is an easier way that we can write this. Just make the multiplicand once with the numeral cards and count how many times we need to make it-three times. Place that many times below the multiplicand.":

$$\begin{array}{r} 2312 \\ \underline{\times \quad 3} \\ 6936 \end{array}$$

3. Explain that there is another way to write it, like a sentence:

3 x 2312 = 6936

Point to the 3 and say, "The three tells us how many times we will take 2312."

**Variation**

Proceed as in Presentation #2, but select a multiplicand and multiplier which will exceed ten in the unit category:

3 x 2314

When all the beads have been combined, begin counting the unit beads until arriving at "ten units." Ask, "What do ten units make?" The sought response is "one ten." Have the child "regroup" or "exchange" the ten units for one ten. Then continue to count the units and get the large numeral card appropriate to the number of unit beads. Continue as in Presentation #l to count the tens, hundreds, and thousands.

The order for introducing "regroup" or "exchange" is:

1. Unit category
2. Ten category
3. Hundred category
4. Unit and hundred
5. Unit and ten
6. Ten and hundred
7. Unit, ten, and hundred

After regrouping or exchanging, introduce a problem that has the product come to zero in the units place, i.e., 2 x 3425. Then proceed to have the ten category come to ten, then hundreds. Then introduce problems in which there are zeros in the multiplicand, i.e., 3 x 2430.

At the end of this sequence, the child should be presented with a random selection of problems.

## Aim

### Direct

Putting together the same group of numbers to understand the special case of addition called multiplication.

Practice in "exchange" or "regrouping."

## Language

| | |
|---|---|
| 2312 | multiplicand |
| x 3 | multiplier |
| 6936 | |

## Age

5 years

# DIVISION WITH THE GOLDEN BEADS (ONE-DIGIT DIVISION)

**Material**

Golden beads, three trays, one large set of decimal numeral cards, three small sets of decimal numeral cards, and four trays.

**Presentation 1**

1. Lay out the set of large numeral cards on a rug and the three sets of small numeral cards to the right.
2. Select a numeral from the large numeral cards and place it on a tray (3693). Select the appropriate quantity of beads to go with the numeral cards.
3. Working with three children, each with a tray, explain that you have a lot of beads that you wish to share with them. But you wish to be fair and give exactly the same amount to each child.
4. Begin to distribute the thousands (we always begin with the largest category), one to each child's tray. When the thousands are distributed, ask, "Does each of you have the same amount?" Then proceed to the hundreds category, since all the thousands have been distributed. Place one of the hundreds on each child's tray, and again ask if the share is equal. Because there are still hundreds remaining, distribute one more hundred to each child. Again, ask if the share is equal. Since there are still hundreds remaining, again distribute one hundred to each child's tray. Ask if the share is equal.
5. Continue in this manner with the tens and the units.
6. Ask each child to count the quantity on his/her tray and select from the small numeral cards the appropriate cards.

7. Then indicate that each child has selected the same numerals because each child has the same quantity. Explain that this special kind of sharing, with each share being the same amount, is called division.
8. When this process has been completed say, "We had 3693 (dividend) and we divided it into three (divisor) equal parts. The amount each child's tray received was 1231 (quotient) and every tray got the same amount."
9. Replace the numeral cards in their respective boxes.

**Presentation 2**

1. Proceed as in Presentation #1. When complete, introduce the division sign: ÷ .
2. Then say, "Since the dividends are all the same, we can write it like a sen tence.":

   3693 ÷ 3 = 1231

   Point to the three and say, "This is the number of equal shares we make."

   You may wish to indicate the other way of writing the problem:

$$\begin{array}{r} \underline{1231} \\ 3\,)\,3693 \end{array}$$

**Variation**

Proceed as in Presentation #2, but select a dividend that will require regrouping or exchange in the thousand category: 4236 ÷ 3. Distribute the thousands to the three trays. There is one thousand left over. We cannot place it on a child's tray because what is placed on one child's tray must be placed on all the trays, and there are not enough thousands. Therefore, the thousand must be regrouped or exchanged for ten of the hundreds. The total of twelve hundreds is then equally distributed among the three trays as are the tens and the units.

The order for introducing "regrouping" or "exchanging" is:

1. Thousands category
2. Hundreds category
3. Tens category
4. Thousands and tens
5. Thousands and hundreds
6. Hundreds and tens
7. Thousands, hundreds, and tens

At the end of this sequence, the child should be presented with a random selection of problems.

By this point, the child has had experience in regroup or exchange and may quickly understand the principle; exchange in any of the categories would be appropriate.

Then introduce problems that have a remainder (R =  ), which is the amount left over after the dividend has been shared equally.

Then introduce problems in which there are zeros in the dividend: 8064 ÷ 4.

**Aim**

**Direct**

Development of order, concentration, coordination, independence and exactness.

Sharing a large quantity equally to understand the concept of division.

Practice in regroup or exchange, going from a larger category to a smaller category.

**Language**

```
  3693  dividend
÷    3  divisor
  ----
  1231  quotient
```

**Age**

6 to 7 years.

**NOTE:**

With the golden bead material, there is a distribution of a quantity into two or more equal parts. This is called distributive division.

# DIVISION WITH THE GOLDEN BEADS (TWO-DIGIT DIVISOR "DECURIAN DIVISION")

**Material**

Golden beads, one set of large decimal cards, one set of small decimal cards, green ribbon to represent unit divisors, blue ribbon to represent tens divisor.

**Presentation 1**

1. Lay out the set of large numeral cards on a rug and the set of small numeral cards.
2. Then call twelve children to the group. Explain that you wish to share a large quantity of golden beads with them. Because this is division, each child must get the same amount.
3. Select a numeral from the large numeral cards (2544), place the cards in a tray, and have a child select the appropriate quantity of beads to go with the numeral cards.
4. Working with three children, each with a tray, explain that you have a lot of beads that you wish to share with tem. But you wish to be fair and give exactly the same amount to each child.
5. Begin to distribute the thousands (we always begin with the largest category), one to each child's tray. When the thousands are distributed, ask, "Does each of you have the same amount?" Since they do not, regroup the 1 thousand to 10 hundreds. Then proceed to the hundreds category, since all the thousands have been distributed. Place one of the hundreds on each child's tray, and again ask if the share is equal. Because there are still hundreds remaining,

distribute the remaining 2 hundreds. Again, ask if the share is equal. Since it is not, regroup the 2 hundreds to 20 tens.

6. Continue in this manner with the tens and the units.
7. Ask each child to count the quantity on his/her tray and say the amount orally.
8. Emphasize that each child now has the same amount.
9. When this process has been completed say, "We had 2544 (dividend) and we divided it into twelve (divisor) equal parts. The amount each child's tray received was 212 (quotient) and every tray got the same amount."

**Presentation 2**

1. Lay out the set of large numeral cards on a rug and the set of small numeral cards.
2. Then call twelve children to the group. Explain that you wish to share a large quantity of golden beads with them. Because this is division, each child must get the same amount.
3. Select a numeral from the large numeral cards (2849), place the cards in a tray, and have a child select the appropriate quantity of beads to go with the numeral cards.
4. Because giving out the beads to so many children took such a long time, a leader will be selected to represent ten of the children. That child will wear the blue ribbon, because blue represents the tens. The other two children will have green ribbons to represent units.
5. The children with the ribbons have trays. Explain that because the leader of the ten children represents ten children, that child must receive ten more than the unit children each time we share a quantity.
6. Beginning with the largest category (thousands), distribute the beads: place a thousand on the tray of the leader and explain that there are ten hundreds in a

thousand, so each of the leader's children will get a hundred. Then place one hundred in each of the unit division trays. They get a share for one.

7. Repeat the above process for the second thousand.
8. Then no thousands remain. Select one of the hundreds and explain that there are ten tens in a hundred. So, the leader for the ten children gives each of the children one ten. Place one ten in each of the unit divisors' trays.
9. Continue with this process until all beads have been distributed, with each tray getting an equal share, and the leader of the ten children getting a share of ten. It will be necessary to regroup a hundred for ten of the ten bars.
10. Because the answer is the "share of one," we look to the tray of the unit divisors to determine what the answer to the problem is.
11. Lay out the large cards. 2844 ÷ 12 = 237
12. So that the children can see that all twelve children really shall receive 237 beads, the leader must proceed to distribute the beads from his/her tray to the ten children. This will involve regrouping. Later, this distribution shall not be necessary as the children realize the answer is "the share of one."

**Aim**

**Direct**

Development of order, concentration, coordination, independence and exactness.

Sharing a quantity with a two-digit divisor underscores that the answer in division is the "share of one."

**Language**

2866 dividend

÷ 23 divisor

237 quotient

**Age**

6 to 7 years.

**NOTE:**

Trays with colored mats can be used to indicate category. Centurion division may be done with a red ribbon. The larger the divisor the shorter will be the process.

# INTRODUCTION TO THE STAMP GAME

**Material**

A tray on which is placed, from left to right, one thousand-cube of golden beads, one hundred square of golden beads, one ten-bar of golden beads, and one golden bead in a caster. One thousand-stamp, one hundred-stamp, one-ten stamp, and one unit-stamp placed beneath the appropriate golden bead category on the tray.

**Presentation**

1. Review the names of the golden bead categories: units, tens, hundreds, and thousands.
2. Introduce the stamps saying, "This green square unit with a one printed on it represents one unit bead." Point to the stamp and the unit bead while saying this for emphasis.
3. Continue with each of the categories of the stamps: tens, hundreds, and thousands. All four categories can be introduced since this is not altogether new for the child.
4. Continue with the second period of the Seguin Three Period Lesson, requesting that the child replace the appropriate stamp with the golden bead category after mixing the order of the stamps.
5. Check the child's comprehension with the third period of Seguin's Three Period Lesson.

**Aim**

**Direct**

Learn the names of the stamp categories and their relationship to the golden beads.

**Indirect**

Prepare to compose quantities with the stamps.

**Age**

5 to 6 years.

**NOTE:**

Originally this material was made of paper stamps which were pasted into notebooks for computation. This proved messy, and the wooden or plastic squares used today retain the anachronistic name, "stamp game." This material is introduced at the point when the child has genuinely tired of the cumbersome manipulation of the golden beads and decimal numeral cards. This is a signal that the child is prepared for a step towards abstraction by using the stamps which represent the quantity. Paper and pencil are used for problems, not the decimal numeral cards. Also, while the golden bead operations may be used with several children working together, the stamp game is an individual activity, usually done at a table.

# FORMATION OF QUANTITY IN STAMPS

**Material**

A box of stamps with green unit squares with the numeral one printed on them, blue ten squares with the numeral ten printed on them, red hundred squares with the numeral one hundred printed on them, and green thousand squares with the numeral one thousand printed on them. The box has six compartments, four of the compartments filled with ones, tens, hundreds, and thousands, respectively. The remaining two compartments should be emptied of the colored counters and colored skittles. Prewritten numerals on slips of paper: 2346, 5247, etc.

**Presentation**

1. Write a numeral or select a numeral from pre-made slips.
2. Place the numeral on the table, select the quantity of stamps, and place the stamps in rows under the numeral, beginning with the units.
3. When completed, read the numeral and "read" the stamps to verify accuracy.
4. Continue with many examples of numerals to compose with stamps for the child to practice independently.

**Aim**

**Direct**

Development of order, concentration, coordination, independence and exactness.

To compose quantities using the stamp material.

**Indirect**

To prepare for the composition of stamps for use in operations.

**Age**

5 to 6 years.

# ADDITION WITH THE STAMP GAME

**Material**

A box of the stamp game materials and prepared addition problems.

**Presentation**

1. Select a problem:

$$\begin{array}{r} 2436 \\ \underline{+1243} \end{array}$$

2. Compose with stamps the two addends, allowing space between the addends but lining up the categories. Then say, "Let's add these amounts together by pushing up all the units together, all the tens together, all the hundreds together, and all the thousands together."
3. Then count the quantity of unit stamps and record the amount on the slip of paper. Continue with the tens, the hundreds, and the thousands, counting and recording each category.
4. Then read the whole problem. 2346 plus 1243 equals 3679.

**Variation 1 (Regrouping)**

Proceed as in Presentation #1, but add numerals in which the unit category will exceed ten: 1436 + 4327. When all the stamps have been combined, begin counting the unit stamps until arriving at "ten units." Ask, "What do ten units make?" "One ten." Have the child "regroup" or "exchange" the ten units for one ten. Then continue to count the unit stamps and record the sum on the paper. Continue as in Presentation #l to count the tens, hundreds, and thousands. The order for introducing "regrouping" or "exchanging" is as follows, if the child continues to be unsure of this process:

1. Unit category

2. Ten category
3. Hundred category
4. Unit and ten
5. Unit and hundred
6. Ten and hundred
7. Unit, ten, and hundred

After regrouping or exchanging, introduce a problem that adds to ten in the unit category, i.e., 2453 + 4237. Then proceed to have the ten category add up to ten, then hundreds.

Then introduce problems in which there are zeros in the addends, i.e., 2450 + 2342 =.

**Variation 2**

Proceed as above, but have three addends.

**Aim**

**Direct**

The operation of addition is explored with the representational material.

**Indirect**

Preparation for the operation of addition with the more abstract material: the dot game.

**Age**

5 to 6 years.

# SUBTRACTION WITH THE STAMP GAME

### Material

A box of the stamp game materials and prepared subtraction problems.

### Presentation

1. Select a problem: 7648 - 2315.
2. Compose the minuend with the stamps (7648).
3. Then, beginning with the unit category, move down the appropriate stamps from the minuend (7648) to form the subtrahend.
4. Pointing to the problem, restate: "We had 7648, we took away 2315 (point to subtrahend stamps), and now we have 4333 remaining.

### Variation 1 (Regrouping)

Proceed as in Presentation #1, but subtract numerals in which the unit category of the subtrahend is more than the unit category of the minuend, resulting in "regrouping": 4364 - 2136. Compose the minuend with the stamps. From the minuend, subtract the subtrahend beginning with the units. Count, "One unit, two units, three units, four units. We need six units, but there are only four units here. How can I get more units? I can regroup a ten for ten units."

Take a ten stamp, place it to the right of the stamp game box, count out ten unit stamps and place them next to the ten stamp. Place the ten stamp in the box and place the units in the unit category of the minuend. Proceed to count out units to make the subtrahend. Continue with each category, and complete the problem as indicated in Presentation #1. With the process of dynamic subtraction, the sequence for introducing problems is:

1. Exchange or regroup in units.

2. Exchange or regroup in tens.
3. Exchange or regroup in hundreds.
4. Zero is in the units of the minuend.
5. Zero is in the tens of the minuend.
6. Zero is in the hundreds of the minuend.
7. Zero is in the units and tens of the minuend.
8. Zero is in the tens and hundreds of the minuend.
9. Zero is in the units, tens, and hundreds of the minuend.

**Variation 2**

Continue as in Presentation #1, but have two subtrahends (intervening subtrahend).

**Variation 3**

Add the subtrahend and the remainder. The sum should equal the minuend.

**Aim**

**Direct**

The operation of subtraction is explored with the representational material.

**Age**

6 years.

# MULTIPLICATION WITH THE STAMP GAME

## Material

A box of the stamp game material and prepared multiplication problems.

## Presentation 1

1. Select a problem: 3 x 2312.
2. Compose the multiplicand, 2312, three times with stamps.
3. Then, beginning with the unit category, push up all the unit stamps together, all the tens together, all the hundreds together, and all the thousands together.
4. Then count the quantity of unit stamps and record the amount under the unit column of the problem. Continue with the tens, the hundreds, and the thousands, counting and recording each category.
5. Then read the whole problem. 2312 taken three times equals 6936.

## Variation 1 (Regrouping)

Select a multiplicand which will exceed ten in the unit category: 3 x 2314.

When all the stamps have been combined, begin counting the unit stamps until arriving at "ten units." Ask, "What do ten units make?" Have the child regroup the ten units for one ten. Then continue to count the units and record the product of the units. Continue with each category.

The order for introducing regrouping, if the child has difficulty with this concept, is:

1. Unit category
2. Ten category
3. Hundred category
4. Unit and hundred categories

5. Unit, ten, and hundred categories

   Introduce a problem that has the product come to zero in the units, i.e., 2 x 3425. Then proceed to have the ten category come to ten, then hundreds. Then introduce problems in which there are zeros in the multiplicand, i.e., 3 x 2340.

**Presentation 2 (Multiplication with a Two-Digit Multiplier)**

1. Select a problem: 13 x 436.
2. Place thirteen green skittles in a column.
3. Next to each skittle make a set of 436 stamps.
4. When all skittles have a set, 436 stamps, group all the stamps together. Regroup unit, tens, and hundreds as needed.
5. The regrouped stamps represent the product: 5668.
6. Record this sum.

**Aim**

**Direct**

The operation of multiplication is explored with the representational material.

**Age**

6 years.

# DIVISION WITH THE STAMP GAME

**Material**

A box of the stamp game material with the skittles to represent the divisor: green skittles to represent units, red skittles to represent tens, blue skittles to represent hundreds, and large green skittles to represent thousands.

**Presentation 1 (One-Digit Divisor)**

1. Select a problem: 3693 ÷ 3.
2. Compose the dividend with stamps: 3693.
3. Place three small green skittles in a row to represent the divisor.
4. Begin distribution with the largest category.
5. Place one of the thousand stamps beneath each of the green skittles.
6. Then ask, "Does each skittle have the same amount?"
7. Then distribute the hundreds, tens, and units. After each distribution, check to see that each skittle received the same amount as the others.
8. When no more stamps remain in the dividend, the distribution is complete.
9. The answer (or quotient) is the share of one or the amount that one skittle has.
10. Record the answer on the paper.

**Variation 1 (Regrouping)**

Select a dividend which will not evenly distribute in the thousand category: 4236 ÷ 3. Distribute the thousands to the three skittles. There is one thousand remaining. It cannot be placed beneath a skittle because there are not enough to share equally with the other two skittles.

Regroup the thousand stamp for ten hundred stamps, and continue to distribute as in Presentation #1. Record the share of one on the paper. By this point the child has had

experience in regrouping and may quickly understand the principle; regrouping in any category would be appropriate.

If the child has difficulty with regrouping the sequence is:

1. Thousands category
2. Hundreds category
3. Tens category
4. Thousands and tens categories
5. Thousands, hundreds, and tens categories

Then introduce problems that have a remainder (R=     ), which is the amount left over after the dividend has been distributed equally.

**Presentation 2 (Two-Digit Divisor "Decurian Division")**

1. Select a problem: 2829 ÷ 23 =.
2. Compose the dividend with stamps: 2829.
3. Place two blue skittles and three green skittles in a row to represent the divisor. Explain to the child that each blue skittle will stand for ten children. Therefore, the two blue skittles stand for twenty children and three green skittles stand for three children. Hence, twenty-three.
4. Place a thousand-stamp beneath the blue skittle (we always begin with the largest category), and say, "This skittle stands for ten children. One thousand is ten hundreds, so this thousandfftamp represents a hundred for each of ten children." Take another thousand-stamp, place it beneath the blue skittle, and repeat. Then select one hundred-stamp and place it beneath the green skittle saying, "The green skittle stands for one child, so I place one hundred beneath it."
5. Repeat this for the two remaining green skittles. Then ask, "Does each skittle have the same amount for each of the children it represents?"

6. Then select a stamp from the largest category (hundreds) and place it beneath the blue skittle, saying, "This skittle stands for ten children. One hundred is ten tens, so, this hundred-stamp represents a ten for each of ten children." Take another hundred-stamp, place it beneath the blue skittle, and repeat. Then select one ten-stamp, place it beneath the green skittle, and say, "The green skittle stands for one child, so I place one ten beneath it."
7. Repeat this for the two remaining green skittles. Continue with this procedure, always giving ten times the amount to the blue skittle as to the green skittle.
8. Record the answer, stating, "The quotient is the share of one, which is found beneath the green skittle: 2829 ÷ 23 = 123.

Of a more complicated nature is the placement of a zero in the divisor and the dividend.

**Presentation 3 (Division with Zero in the Dividend)**

1. Select a problem: 3036 ÷ 21 =.
2. Compose the dividend with the stamps: 3036.
3. Place two blue skittles and one green skittle to represent the divisor.
4. Place one thousand beneath each blue skittle (remember, we always begin with the largest category). This represents one hundred for each of ten children.
5. Then the green skittle should receive one hundred. Because there are no hundreds, regroup a thousand-stamp for ten hundreds and proceed as in Presentation #2.
6. Record the answer, stating, "The quotient is the share of one, which is found beneath the green skittle." 3036 ÷ 21 = 144 R=12.

**Presentation 4 (Division with Zero in the Divisor)**

1. Select a problem: 7463 ÷ 203.
2. Compose the dividend with stamps: 7463.
3. Place two red skittles, one blue counter, and three green skittles in a row to represent the divisor. The blue counter represents zero tens.
4. Place one thousand beneath each of the red skittles. A hundred stamp would then be placed beneath the blue skittles; however, there are zero, as represented by the blue counter. Therefore, a ten stamp is placed beneath each green skittle.
5. This process is repeated; each time a share is to be given to the tens divisor pass on to the unit divisor, selecting a stamp of one category less.
6. Continue until the divisor has been distributed.
7. Record the answer, stating, "The quotient is the share of one, which is found beneath the green skittle: 7463 ÷ 203 = 36 R=155.

**Presentation 5 (Division with Zero at the Unit of the Divisor)**

1. Select a problem: 8760 ÷120 =.
2. Compose the dividend with stamps: 8760.
3. Place one red skittle, two blue skittles, and one green counter in a row to represent the divisor. The green counter represents zero units.
4. Proceed as in the earlier presentation to distribute the stamps.
5. Upon completion of the distribution, say, "The answer is the share of one. Do we have a share of one?"
6. Point to a blue skittle and say, "These stamps represent ten shares."
7. Proceed to place ten green skittles in a row and distribute one share of ten (the stamps from one green skittle) evenly among the ten skittles.
8. Record the answer, stating, "The quotient is the share of one, which is found

beneath the green skittle." 8760 ÷120 = 173.

Through repeated practice, the child will realize that the last zero of the blue ten skittle's stamps (730) need merely be dropped to give the answer (73).

Presentation 6 (Group Division)

In distributive division, a particular amount is evenly distributed to the divisor.

In group division, the divisor becomes a particular size group into which the dividend is subdivided.

1. Select a problem: 5232 ÷ 436 =.
2. Compose the dividend with stamps: 5232.
3. Do not place skittles out for the divisor because this is not distributive division.
4. Proceed to group the divisor (5232) into sets of 436.
5. Regroup, as necessary, until all stamps have been grouped into sets of 436.
6. Record the answer, stating, "The quotient is the number of sets that have been made." There are 122 sets of 436. Therefore, the quotient is 12.

   5232 ÷ 436 = 12.

Presentation #6 is important for understanding abstract division.

**Aim**

**Direct**

Development of order, concentration, coordination, independence and exactness.

The operation of division is explored with the representational material.

**Age**

7 to 8 years.

# THE DOT BOARD: ADDITION

## Material

A wooden-framed clear plastic board with a matrix sheet beneath, the matrix sheet has six columns and four rows. The columns are labeled from left to right: 10,000, 1,000, 100, 10, 1 in the green, blue, and red color-coding of the decimal system. The sixth column is unlabeled and is used to state the minuends and sums of the problem. The four rows, from top to bottom are: 1) the labels of the category; 2) a grid of ten squares contained in each column of the second row; 3) a grid of ten squares contained in each column of the third row to record regroupings; and 4) the fourth row is blank allowing for totals of the columns. This material is also available in paper.

## Presentation 1

1. In the right-hand column record a problem in addition with multiple addends:

   4362 + 5243 + 1478 + 2369 =

2. Beginning with the first addend, record the amount in dots under the appropriate category heading; begin in the top row and dot from right to left, placing one dot to a square:

A) In the units column, record two green dots beginning from the right side.

B) In the tens column, record six blue dots beginning from the right side.

C) In the hundreds column, record three red dots beginning from the right side.

D) In the thousand column, record four green dots beginning from the right side.

3. Proceed in this manner to record each addend, entering the dots to complete each row before continuing to the next row.

4. When all the addends have been entered, begin the process of regrouping. Use a black pencil for regrouping. Starting with the units, cross out the first row of ten dots; record a dot in the tens column below the grid to the left and

a black dot in the tens matrix. Continue until all complete rows are crossed out and recorded. Record the numeral 2 in the tens column above the grid.

5. Count the number of dots remaining in the units category and record this numeral in the last row of the unit column.
6. Proceed in this manner for each category, regrouping to the next column; record the remaining dots of the category in the final row of the appropriate category.
7. The numerals recorded in the last row represent the sum of the addends. Record this sum under the addends in the right column.

**Presentation 2**

1. In the right column, record a problem in addition with multiple addends:

   4362 + 5243 + 1478 + 2369 = .
2. Dot in the units of all of the addends. Cross out rows which have complete sets of ten unit dots. Record these sets in the tens column as indicated in Presentation #1. Count the number of unit dots remaining and record this numeral in the last row of the unit column.
3. Proceed in this manner for the tens, hundreds, and thousands.
4. The numerals recorded in the last row represent the sum of the addends. Record this sum under the addends in the right column.

**Aim**

**Direct**

Ten of a lower category represents one of a higher category.

The operation of addition is explored with a more abstract material.

**Age**

7 years.

**NOTE:**

Simple multiplication may be performed on the Dot Board.

# INTRODUCTION TO THE BEAD FRAME

The bead frames represent the passage from the real quantity to representative hierarchical quantities. In the golden bead material, the objects which represented 10 - 100 - 1000 were in the form of a bar, square, and cube and were actual quantities 10, 100, or 1000 times the unit.

Now, with the hierarchical material, 10 is represented by one head, as is 100 and 1000. There is, in the bead frame, a logical internal development in which the number of objects no longer is the element which gives value to the number. The position of the bead indicates the value: For example: The blue bead on the second wire is ten times the unit bead and one tenth the value of the hundred bead. We know its value from its position on the frame.

In multiplication, there are two aspects:

1. The memorization of the basic equations. This is done with the memorization material.
2. The rapid localization of the place of the digit of the number. This is done with the bead frame.

With the bead frame, the most important consideration is the position of the digit. That is, the value of the digit is relative to the position it occupies and no longer relative to a quantity of material. Every digit has an absolute value, independent of the position it occupies in a number. It also has a relative value which is dependent on the position it occupies in the number.

To illustrate this fact, Dr. Montessori make an analogy. Let us consider our society: We have a president, cabinet members, judges, congressmen, and simple citizens. Considering them all as men, they are all equal. But what distinguishes them from one other? It is their varying social positions in government. It is the same with the numerals. They can be units or they can be millions. It is the position occupied by the

numeral which gives its value.

With the bead frame, we are confronted with material in which the value of a numeral is in its position and not in quantity. When the child worked with the golden beads, the quantity could be displayed in any position. No matter what order the material had with regard to placement, the child identified the quantity from the beads. Now these quantities are reduced to symbol. Each quantity must have its own place, the place it holds in decimal numeration. Now the beads have place value.

Color is introduced to represent the hierarchies:

Units - green

Tens - blue

Hundreds - red

The child first encounters these colors in the decimal layout cards and in the stamp game.

Before working with the bead frame, the terms "absolute" and "relative" are clarified. Consider in beads: one hundred - twenty - four. No matter their placement, their value is the same.

# INTRODUCTION TO THE BEAD FRAME WITH THE GOLDEN BEADS

**Material**

The Bead Frame. Ten of each of the categories of the golden beads.

**Presentation**

1. Review the golden bead quantities.
2. Present the bead frame.
3. Place the golden beads beside the bead frame.
4. Place a unit-bead in front of the bead frame.
5. Move one green unit-bead from the left to the right.
6. Place a second unit-bead in front of the bead frame.
7. Move a second green unit-bead from the left to the right.
8. Continue in this manner with nine golden unit-beads and nine green unit-beads.
9. Lay out the tenth unit-bead and exchange the unit-beads for a ten-bar.
10. Move the tenth green bead across. Then move the ten beads to the left to the center of the wire.
11. Move a blue ten-bead to the right and to the center of the wire. Indicate their equal value.
12. Continue to move the green unit-beads to the left and the one blue tenbead to the right.
13. Continue with the ten golden bead bars and blue ten-beads.
14. When the tenth golden bead bar is placed down, exchange the ten bead bars for a hundred square.
15. When the tenth blue ten-bead is moved to the right, exchange the ten-beads

for a red hundred-bead.

16. Continue in this manner through to thousands.
17. Upon completion, place a golden bead quantity out and have the child repre sent it on the bead frame.
18. Continue to display golden bead quantities and have the child represent them on the bead frame.

# INTRODUCTION TO THE BEAD FRAME WITH NOTATION PAPER

**Material**

A bead frame with four rows. Notation paper for the bead frame.

**Presentation**

1. Point to the notation paper and say, "I shall use this paper to record the quantity indicated on the bead frame."
2. Move one unit-bead from the left side of the frame to the right side of the frame and say, "One."
3. Record the numeral one on the notation paper by writing a one on the upper right-hand line and through the vertical line.
4. Move a second unit-bead from the left side of the frame to the right side of the frame and say, "Two."
5. Record the numeral two on the notation paper by writing a two on the line below the one and through the vertical line.
6. Continue in this manner through the ninth bead, noting the numeral on paper each time.
7. Move the tenth unit-bead from the left side of the frame to the right side of the frame and say, "Ten."
8. Move the ten unit-beads from the right to the center of the wire.
9. Move one ten-bead from the left to the center of the wire.
10. Point to the ten units and say, "Ten." Point to the one ten-bead and say, "Ten."
11. Move the unit-beads to the left.
12. Move the ten-bead to the right.
13. Record the numeral one in the tens column with no zero on the notation

paper.

14. Continue to count the ten-beads and record them on the notation paper with out zeros.
15. Move the tenth ten-bead from the left side of the frame to the right side of the frame and say, "One hundred."
16. Move the ten ten-beads from the right to the center of the wire.
17. Move one hundred-bead from the left to the center of the wire.
18. Point to the ten tens and say, "One hundred." Point to the one hundred-bead and say, "One hundred."
19. Move the ten-beads to the left.
20. Move the hundred-bead to the right.
21. Record the numeral one in the hundred column with no zero on the notation paper.
22. Continue to count hundred-beads and record them on the notation paper with out the zeros.
23. Move the tenth hundred-bead from the left side of the frame to the right side of the frame and say, "One thousand."
24. Move the ten hundred-beads from the right to the center of the wire.
25. Move one thousand-bead from the left to the center of the wire.
26. Point to the ten hundreds and say, "One thousand."
27. Move the ten hundred-beads to the left.
28. Move the thousand-bead to the right.
29. Record the numeral one in the thousand column on the notation paper.
30. When all the notation is completed to 1000, fill in the zeros with a red colored pencil. Adding the necessary number of zeros at the end impresses on the child'3 mind the number of zeros needed for the various categories.

**NOTE:**

The printed form reproduces the bead frame twice. The dotted line separating the hundreds from units of thousands represents the comma which is used in notation of four-digit numbers.

# READING AND WRITING NUMBERS ON THE BEAD FRAME

**Material**

Bead frame and bead frame notation paper.

**Presentation**

1. The teacher forms the quantity with beads and asks the child to name it and write it. First an easy quantity is formed; later the zero is introduced in various positions.
2. The teacher writes a number. The child reads it and forms the quantity in beads on the frame.
3. The child names the quantity and writes the number on the notation paper.
4. Orally, ask the child to make quantities on the bead frame.

**Aim**

**Direct**

Development of order, concentration, coordination, independence and exactness.

To constantly bring to the child's consciousness the function of position, i.e., place value.

# ADDITION WITH THE BEAD FRAME

**Material**

The bead frame, special paper for bead frame notation, and pencil.

**Presentation 1**

1. Record a number on the notation paper, placing the numerals on the proper colored lines. (2436)
2. Represent this numeral with beads on the bead frame.
3. Enter a second number on the paper to add to the first quantity (1243).
4. Represent this numeral with beads on the bead frame.
5. The sum is represented by the quantity of beads found to the right side of the bead frame.
6. Beginning with the units, record the sum on the notation paper. (3679)

**Presentation 2 (Regrouping)**

Proceed as in Presentation #1, but this time select a problem in which the units shall exceed ten when added. Regrouping the ten units for a blue ten-bead will be necessary. Then proceed to continue to count out the units. Continue with addends that require regrouping in the tens, hundreds, and thousands categories.

**Aim**

**Direct**

The operation of addition is explored with this most abstract material.

**Age**

6 to 7 years.

# SUBTRACTION WITH THE BEAD FRAME

**Material**

Bead Frame, special bead frame notation paper, and a pencil.

**Presentation 1**

1. Notate a numeral on the paper: 7648.
2. Compose the minuend on the bead frame.
3. Then notate the subtrahend: 2315.
4. Using the units to the right side of the frame (the minuend), move the units of the subtrahend back to the left. Continue in this manner with the tens, hundreds, and thousands.
5. The beads remaining to the right side of the bead frame represent the remainder.
6. Record the remainder on the notation paper.

**Presentation 2**

Proceed as in Presentation #1, but subtract numerals in which the unit category of the subtrahend is more than the unit category of the minuend, resulting in regrouping:

4364 - 2136.

Compose the minuend. Move the four unit-beads to the left. Two more are required. Therefore, regroup by moving one ten-bead from the minuend (right side) to the left and moving ten unit beads from the left to the right. Proceed to subtract the two more unit-beads. Then continue with the subtraction. The quantity remaining to the right represents the remainder (2228). Record it on the notation paper.

**Aim**

**Direct**

Development of order, concentration, coordination, independence and exactness.

The operation of subtraction is explored with this most abstract material.

**Age**

6 to 7 years.

# CHAPTER IV

# MEMORIZATION

# INTRODUCTION TO MEMORIZATION WORK

The child has first experienced the operations with the golden beads. The concepts of addition, subtraction, multiplication, and division are understood. However, as the child moves through the more representational and abstract materials of the stamp game, the dot game, and the bead frame, there is the need for the child to commit the facts to memory in order to be freed from the concrete manipulation of materials.

The Webster's Collegiate Dictionary defines "memorize" as, "to commit to memory" and "to learn by heart." The following activities aid the child in the process of memorizing the addition, subtraction, multiplication, and division facts. In each of these operations there is a series of activities which repeats the equations so as to commit the facts to the child's memory in an interesting and challenging way.

It is possible to commit facts to memory in a rote way, but this is boring for the child and in no way strikes the child's imagination. The Montessori sequence of materials holds the child's interest as points of interest are revealed.

So, in addition to presenting memory work in a highly interesting and imaginative context, the materials increasingly reveal sophisticated aspects of each operation:

1. In addition and multiplication, the commutative law is illustrated in one of the exercises.
2. In all the operations, the factoring from a sum remainder and product are explored.
3. In all the operations the child is led to discover the essential facts needed to memorize.

Therefore, while the direct aim of the materials is to lead the child to the memorization of the facts, the child is led through these activities to experience increasingly more sophisticated math concepts. These lay the foundation for an "intuitive" grasp of algebra later.

# INTRODUCTION TO THE ADDITION STRIP BOARD

**Material**

A board divided into eighteen 2-cm. squares across and twelve 2-cm. squares down. The vertical column is numbered one through eighteen: one through ten in red, and eleven through eighteen in blue. A red vertical line after the ten column divides the board.

Strips of wood: One set of nine blue strips in increasing lengths (from two to eighteen cm.), with a red num eral on the right end of each strip. One set of nine red strips in increasing lengths (from two to eighteen cm.) with blue vertical lines dividing each strip into 2-cm. segments. The numerals one through nine are printed in blue at the right end of the strips.

**Presentation**

1. Place the board in the middle of the table. Place the blue strips above the board on the left side in sequence from nine through one, top to bottom. Place the red strips above the board on the right side in sequence from nine through one, top to bottom.
2. Select a blue strip and place it on the top row of the board, having left edge of the strip flush with the left side of the board. Select a red strip and place it next to the blue strip. The answer to this addition appears over the last segment of the red strip.
3. Several more examples may be used to illustrate the use of the addition strip board.
4. Allow the child to create addition problems, noting their sums.

**Aim**

**Direct**

To acquaint the child with the use of the addition strip board.

**Age**

5 years.

**NOTE:**

The purpose of the vertical red line down the strip board after the ten column is to emphasize sums of ten. It serves the same function as the notched card in the snake game: to separate a group of ten from a larger number.

# ADDITION STRIP BOARD

**Material**

The addition strip board

Strips

Booklet of prepared equations

First control chart

**Presentation 1 (Booklet)**

1. Refer to the first page of the booklet, which has the 1 + table.
2. Read the first equation: 1 + 1 =
3. Place the blue one-strip on the first row of the board. Place the red one-strip on the first row of the board to the right of the blue strip. Read 1 + 1 = 2, pointing to the answer above the end of the red strip.
4. Record the sum on the line provided in the booklet.
5. Replace the red strip. The blue stays stationary for it indicates that the table is being worked on.
6. Read the second equation: 1 + 2 =
7. Place the red two on the first row of the board to the right of the blue strip. Read 1 + 2 = 3, pointing to the answer above the end of the red strip.
8. Record the sum on the line provided in the booklet.
9. Replace the red strip.
10. Proceed with the following equations until the page is complete.
11. Then continue through several more pages.
12. Refer to the Addition Control Chart 1 to check accuracy.
13. Continue through the booklet.

**Aim**

**Direct**

To acquaint the child with the basic addition sums.

**Age**

5 to 6 years.

**NOTE:**

This activity will not hold the attention of the elementary age child for long.

**Presentation 2 (Prepared Equation Slips, First Control Chart, and Paper for Recording)**

1. The child selects an equation and records it in his math notebook.
2. The blue strip is selected for the first addend and placed on the board. The red strip is selected for the second addend. Their sum is noted and recorded on the paper.
3. The equation slip is placed to the side.
4. The child continues in this manner.
5. Upon completion, the child refers to the Addition Control Chart to check the accuracy of the sums.

**Aim**

**Direct**

To aid the child in the memorization of the addition facts.

**Age**

5 to 6 years.

**Presentation 3 (Combinations of Numbers: First and Second Control Charts and Paper for Recording Work)**

1. Say, "Let's see how many ways we can make ten."
2. Select the blue one-strip and the red nine and place them on the board.
3. Record the equation formed: 1 + 9 = 10.
4. The strips remain on the board.
5. Select the blue two and the red eight and place them on the board directly under the 1 + 9.
6. Record the equation formed: 2 + 8 = 10.
7. The strips remain on the board.
8. Proceed in this manner to 9 + 1 = 10.
9. Note the pattern established with the increase of the blue strip and the decrease of the red strip. (The child may remember parallel work in the number rods.)
10. Indicate the 1 + 9 = 10 and 9 + 1 = 10 equations. Say, "One plus nine equals ten; nine plus one equals ten. The numerals one and nine may be added either way and still equal ten. Therefore, it is only necessary to remember 1 + 9 = 10." Then turn over the 9 + 1 strips at the bottom (Beroni) or replace the strips above the board (Nienhuis). Draw a line through the written equation 9 + 1 = 10.
11. Continue with 2 + 9 and 8 + 2, 3 + 7 and 7 + 3, and 4 + 6 and 6 + 4.
12. At 5 + 5 indicate that there is not another equation that duplicates 5 + 5. Therefore, nothing may be deleted.
13. Note that by this process of elimination only five equations need be memorized, not the nine originally laid on the board.
14. Analyze Control Chart 1 with the same sums running diagonally from bottom to top, left to right. Indicate the diagonal for sums of ten.

15. Then proceed to eliminate the duplicate equations, beginning at the top left of the chart. 1 + 1 = 2 may not be eliminated, for there is not duplicate.
16. Then analyze the sum 3. It is found in two equations: 2 + 1 = 3 and 1 + 2 = 3. Therefore, one of the equations may be eliminated. Place a blank, rectangular paper over 2 + 1 = 3.
17. Continue to analyze the sums to eighteen, eliminating duplicate equations.
18. An interesting pattern emerges as the number of equations for a particular sum increases to a peak at ten and then decreases at the same rate of increase to only one equation to analyze at sum 18, as there was only one equation to analyze for the sum of 1. Following this same pattern is the number of duplicate equations to be analyzed.
19. At the end of the analysis of Control Chart 1, the child sees that nearly half the chart has been covered, reducing the total number of equations to be memorized from 81 to 45.
20. Compare Control Chart 1 and Control Chart 2; it is apparent that Chart 2 contains only the equations needed. Chart 2 has been arranged to have equal sums fall on the same horizontal line.

## Aim

### Direct

Development of order, concentration, coordination, independence and exactness.

To bring to the child's consciousness the commutative law of addition.

## Age

5 to 6 years.

**NOTE:**

Sums dealing with 1 - 9 are explored here. Sums 10 - 18 follow the same procedure.

**Presentation 4 (Combinations of Numbers with Zero)**

1. Say, "Let's see what happens when we add 0."
2. Record the equation 0 + 9 = ___.
3. Place the red nine strip on the board. Say, "Zero plus nine equals nine." Record the answer.
4. Continue with the equations that have 0 as an addend. 0 + 6, 0 + 5, 0 + 4, 0 + 3, 0 + 2, 0 + 1, 0 + 0.
5. What happens when we add zero?

**Aim**

**Direct**

To aid the child in the memorization of the addition facts.

**Age**

5 to 6 years.

**Presentation 5 (Doubles of Numbers: Control Chart 2)**

1. Say, "Let's add the same number twice."
2. Select the blue one-strip and the red one-strip and place them on the board.
3. Note that 1 + 1 = 2.
4. Record the equation.
5. Replace the blue one-strip and the red one strip.
6. Proceed with the doubles: 2 + 2 =, 3 + 3 =, 4 + 4 =, 5 + 5 =, 6 + 6 =, 7 + 7 =,

8 + 8 =, 9 + 9 =.

7. Refer to Control Chart 2 to check for accuracy. Note that the doubles lie on a diagonal that extends from top to bottom, left to right.

**Aim**

**Direct**

Development of order, concentration, coordination, independence and exactness.

To aid the child in memorizing the addition facts.

To explore the addition of two equal addends.

**Age**

6 years.

# ADDITION CHART 3

## Material

Addition Chart 3

Prepared equation slips

Paper and Pencil

Control Chart 1

## Presentation 1

1. Select a prepared equation slip.
2. Record the equation.
3. The sum is found by placing an index finger on the blue numeral of one addend and the other index finger on the red num eral of the other addend. Bring one finger down the column and bring the other finger across the row. At the square where they meet is the sum.
4. Record the sum.
5. Continue in this manner.
6. Upon completion, refer to Control Chart 1 to check accuracy.

## Aim

### Direct

Development of order, concentration, coordination, independence and exactness.

To aid the child in the memorization of the addition facts.

## Age

6 years.

**Presentation 2**

1. Indicate that the sum of the doubles are found in the diagonal squares running from top to bottom, left to right. It divides the chart in half. There are the same number of sums above the diagonal as below the diagonal and in the same position.
2. Then analyze the sums for duplicate equations. There is only one sum 2. The equation for it cannot be eliminated.
3. The sum 3 is found twice: 2 + 1, and 1 + 2. They are duplicates so one may be covered with a blank square over the sum 3.
4. Continue to analyze the sums to eighteen, eliminating duplicate equations.
5. Upon completion, compare Chart 3 with Chart 4. They are the same, with Chart 4 containing only the unduplicated sums. The child has created it and understands its rationale. (The blue addends at the top are missing on Chart 4.)
6. Ask the child how many sums remain on Chart 3 after duplicates have been eliminated (45).

**Aim**

**Direct**

To reduce the number of equations to be memorized.

**Age**

6 years.

**NOTE:**

Chart 3 is similar to Control Chart 1 in information and organization. However, Chart 3 is more abstract because it contains the sum in each square, not the whole equation.

The addends are formed with the fingers. This process of elimination requires a higher level of abstraction than on Control Chart 1 because the child must begin with the sum and find the addends by moving fingers up and over.

Charts 3, 4, and 5 are sometimes referred to as "finger charts" for obvious reasons.

# ADDITION CHART 4 (THE HALF TABLE)

## Material

Addition Chart 4

Prepared equation slips

Paper and pencil

Control Chart 1

## Presentation

1. Select a prepared equation slip.
2. Record the equation.
3. Place one index finger on one addend in the red column, and the other index finger on the other addend in the red column.
4. Slide both fingers over to the right as far as possible.
5. Then move the top finger in a down-over movement, one square at a time, while moving the bottom finger in an over-up movement, one square at a time at the same rate as the top finger. Where they meet is the sum.
6. Record the sum.
7. Proceed in this manner, selecting prepared equation slips.
8. Refer to Addition Control Chart 1 to check accuracy.

## Aim

### Direct

To aid the child in the memorization of the addition facts.

## Age

6 years.

**NOTE:**

If both addends are even, the sum is even and, therefore, on the outer edge of the chart. If both addends are odd, the sum is even and, therefore, on the outer edge of the chart. If one addend is even and the other addend is odd, the sum is odd and along the second diagonal from the top. (The even sums are on the top diagonal; the odd sums are on the diagonal below it. This is why it is important to demonstrate an even sum and an odd sum equation.)

# ADDITION CHART 5 (SIMPLIFIED CHART)

**Material**

Chart 5

Prepared equation slips

Paper and pencil

Control Chart 1

**Presentation**

1. Select a prepared equation slip.
2. Record the equation.
3. Place one index finger on one addend in the red column, and the other index finger on the other addend in the red column.
4. Slide both fingers over to the right as far as possible.
5. Then move the top finger in a down-over movement, one square at a time, while moving the bottom finger in an up-over movement, one square at a time, at the same rate as the top finger. Where they meet is the sum.
6. Record the sum.
7. Proceed in this manner, selecting prepared equation slips.
8. Refer to Addition Control Chart 1 to check accuracy.
9. Analyze the addends and sums for:

    even + even = even

    odd + odd = even

    even + odd = odd.

## Aim

### Direct

To aid the child in the memorization of the addition facts.

## Age

6 years.

# ADDITION CHART 6 (BLANK CHART)

**Material**

Addition Chart 6 (Blank Chart)

Stamps with sums

Prepared equation slips

Control Charts 1 and 3

Math notebook

**Presentation 1 (Bingo Game #1)**

1. Lay all the stamps face up on the table.
2. Select an equation and record it in the math notebook.
3. From the stamps, select the correct sum and place it at the top of the Addition Chart 6.
4. With the index finger of the right hand, find one addend along the top of Addition Chart 6.
5. With the index finger of the left hand, find the other addend along the left side.
6. Move the right index finger down and the left index finger across. Where they meet is the blank square where the sum stamp should be placed.
7. Record the sum in the math notebook.
8. Select another prepared equation slip and proceed as before.
9. Upon completion of Addition Chart 6, compare it to Addition Chart 3. They should be identical. Compare the equations in the math notebook to Addition Chart 1 for accuracy.

## Aim

### Direct

To memorize addition facts.

## Age

6 to 7 years.

## Presentation 2 ("Bingo Game #2")

1. Lay all the stamps face up on the table.
2. Select a sum (stamp) at random and place it at the top of Addition Chart 6. Record it in the math notebook.
3. The child makes up an equation to satisfy the sum and enters the complete equation in the math notebook.
4. With the index finger of the right hand, find the first addend along the top of the Addition Chart 6.
5. With the index finger of the left hand, find the other addend along the left side.
6. Move the right index finger down and the left index finger across. Where they meet is the blank square where the sum stamp should be placed.
7. Select another remainder (stamp) and proceed as before.
8. Upon completion of Addition Chart 6, compare it to Addition Chart 3. They should be identical. Compare the equations in the math notebook to Addition Chart 1 for accuracy.

## Aim

### Direct

To memorize the addition facts.

**Age**

6 to 7 years.

**Presentation 3 ("Bingo Game #3")**

1. The child lays all the stamps face up in stacks of the same denomination from 1 through 9.
2. Select a sum stack (stamps).
3. Make up the possible addition equations to satisfy the sum. Each equation is entered in the math notebook, and with the index fingers, the position of the sum on the Addition Chart 6 is located.
4. Upon completion of the stacks of sums (stamps), compare the filled-in Chart 6 (blank chart) with Chart 3. They should be identical. Compare the equations in the math notebook to Addition Control Chart 1 for accuracy.

**Aim**

**Direct**

Development of order, concentration, coordination, independence and exactness.

To memorize the addition facts.

**Age**

6 to 7 years.

# ORAL GAMES FOR THE MEMORIZATION OF ADDITION

### Material

Addition equation slips

Sum stamps

### Presentation 1

1. The child selects an addition equation slip and reads it.
2. Another child says the sum.
3. The first child determines the accuracy of the response from memory or from Control Chart 1.
4. Proceed with the other equation slips in this manner.

### Aim

#### Direct

To memorize the addition facts.

### Age

6 to 7 years.

### Presentation 2

1. The child selects a sum (stamp) and reads it.
2. Another child says the possible equations to satisfy this remainder.
3. The first child determines the accuracy of the responses from memory or from Control Chart 1.
4. Proceed with the other sum stamps in this manner.

**Aim**

**Direct**

Development of order, concentration, coordination, independence and exactness.

To memorize the addition facts.

**Age**

6 to 7 years.

# ADDITION SNAKE GAME

**Material**

A box containing colored bead bars for the numerals one to nine. There should be eight to ten bars for each number.

One - red

Two - green

Three - pink

Four - yellow

Five - blue

Six - lavender

Seven - white

Eight - brown

Nine - dark blue

A box with a set of black and white bead bars. All the bead bars through five are black. From six through nine the bead bars are black for the first five beads and white for successive beads. This allows for ease in identifying the quantity on the bead bar.

A box of ten bead bars

A notched strip of cardboard

A box of golden bead bars

**Presentation**

1. Remove the black and white bead bars and place them to the upper right of the work area in graded order.
2. Place the box of golden bead bars in the upper left of the work area.
3. Select an arrangement of colored bead bars and lay them "snake-wise" across the work area.

4. Tell the child to take the first two bars and move them away from the snake. The beads should be added without counting: 4 + 7 = 11.

5. Place a golden ten and the black one in the snake.

6. Move down the next two bars: 1 + 8 = 9.

7. Exchange the one black bead and the eight bead for a black and white nine.

8. Continue in the same way to the end of the snake, each time adding the black or black and white bead to the next colored bead, add from memory, then exchange the beads for golden beads and/or black and white beads.

9. Verification: "Let's see if I have added correctly." Line up the ten bars to the left. Line up the colored bead bars to the right, grouping like colors together. Next to each ten place a combination of colored bars that will add up to ten. When a one-to-one correspondence has been made between the colored bead bars and the ten bars, the verification is complete.

**Aim**

**Direct**

To aid the child in the memorization of the addition facts.

**Age**

6 years.

**NOTE:**

When adding any set of beads, if the sum is not ten then the black and white beads must be used. If the black and white beads are used then the next set of beads to be added will contain the black and white bar and a colored bead bar.

# BEAD BARS FOR THE MEMORIZATION OF ADDITION

**Material**

A box with ten compartments containing twenty bead bars in each compartment 1 - 10.

Labels with + and = signs.

**Presentation 1**

1. Select a prepared slip with an addition equation that will sum to a total of less than ten: 4 + 3 =.
2. Select the appropriate bead bars to represent this equation and place them horizontally with the + and = labels placed appropriately between the bead bars.
3. Sum this quantity from memory. Select the appropriate beads to represent the sum.
4. Place the sum beads vertically after the = sign.
5. Record this equation in the child's math notebook.
6. The child can continue with the prepared slips.

**Aim**

**Direct**

To aid in the memorization of the addition facts.

**Age**

6 years.

**Presentation 2**

1. Select a prepared slip with an addition equation that will sum to a total of more than ten: 4 + 9 =.
2. Select the appropriate bead bars to represent this equation and place them horizontally with the + and = labels placed appropriately between the bead bars.
3. Sum this quantity from memory. Select the appropriate beads to represent the sum.
4. Place the sum beads vertically after the = sign.
5. Record this equation in the child's math notebook.
6. The child can continue with the prepared slips.

**Aim**

**Direct**

To aid in the memorization of the addition facts.

**Age**

6 years.

# BEAD BARS: COMMUTATIVE LAW

**Material**

A box with ten compartments containing twenty bead bars in each compartment 1 - 10.

Labels with + and =.

**Presentation**

1. Prepare two equations and two slips: 6 + 7 =, 7 + 6 =.
2. Working with one equation ( 6 + 7 =) at a time, select the appropriate bead bars to represent this equation and place them horizontally with the + and = labels placed appropriately between the bead bars.
3. Sum this quantity from memory. Select the appropriate beads to represent the sum.
4. Place the sum beads vertically after the = sign.
5. Record this equation in the child's math notebook.
6. Select the second equation (7 + 6 =) and repeat the process.
7. Continue with all the possible addition equations and place them in pairs that reflect the commutative law of addition.

**NOTE:**

Do not tell the child about the commutative addends. When the child realizes the relationship, then give the "law."

**Aim**

**Direct**

Development of order, concentration, coordination, independence

and exactness.

To bring to the child's consciousness the commutative law of addition.

**Age**

6 years.

# BEAD BARS: MULTIPLE ADDENDS

## Material

A box with ten compartments containing twenty bead bars in each compartment 1 - 10.

Labels with + and =.

## Presentation

1. Select a prepared slip with an addition equation with several addends: 2 + 1 + 3 + 4 =.
2. Select the appropriate bead bars to represent this equation and place them horizontally with the + and = labels placed appropriately between the bead bars.
3. Sum this quantity from memory. Select the appropriate beads to represent the sum.
4. Place the sum beads vertically after the = sign.
5. Record this equation in the child's math notebook.
6. The child can continue with the prepared slip.

## Aim

### Direct

Development of order, concentration, coordination, independence and exactness.

To bring to the child's awareness that addition can continue on with many addends.

## Age

6 years.

# BEAD BARS: ADDENDS LARGER THAN TEN

**Material**

A box with ten compartments containing twenty bead bars in each compartment 1 - 10.

Labels with + and =.

**Presentation 1**

1. Select a prepared slip with an addition equation of two two-digit addends and involving no exchange: 13 + 25 =.
2. Select the appropriate bead bars to represent this equation and place them vertically with the + and = labels placed appropriately between the bead bars.
3. Sum the addends beginning with the units place. Place the beads for the sum vertically to the right of the = sign.
4. Record this equation in the child's math notebook: 13 + 25 = 38.
5. The child may continue with the prepared slips.

**Aim**

**Direct**

To use the addition facts in more than the unit category.

**Age**

6 years.

**Presentation 2**

1. Select a prepared slip with an addition equation of two two-digit addends and

involving exchange in the units: 16 + 27 =.

2. Select the appropriate bead bars to represent this equation and place them vertically with the + and = labels placed appropriately between the bead bars.
3. Sum the addends beginning with the units place. The sum will be more than nine, so the sum of the units will require representation with a ten bead-bar. Place the beads for the sum of the units vertically to the right of the = sign, but hold the ten bead-bar in your hand.
4. Sum the tens, adding in the ten-bead in your hand, and place the beads vertically to the right of the = sign.
5. Record this equation in the child's math notebook: 16 + 27 - 43.
6. The child may continue with the prepared slips.

**NOTE:**

Later, in Presentation #2, have the child hold in memory the ten-bar that is exchanged in the units.

**Aim**

**Direct**

Development of order, concentration, coordination, independence and exactness.

To use the addition facts in more than the unit category.

**Age**

6 years.

# BEAD BARS: ASSOCIATIVE LAW

**Material**

A box with ten compartments containing twenty bead bars in each compartment 1 - 10.

Labels with + and =.

**Presentation 1**

1. Select a prepared slip with an addition equation of three addends: 3 + 5 + 7 =.
2. Select the appropriate bead bars to represent this equation and place them horizontally with the + and = labels placed appropriately between the bead bars.
3. Bring the first two addends down (3 + 5) and add them.
4. Replace the three bead-bar and the five bead-bar with the eight bead-bar.
5. Remove the extra + sign.
6. Sum the remaining two addends and place the sum beads vertically after the = sign.
7. Record this equation in the child's math notebook:

   3 + 5 + 7 = 15

   8 + 7 = 15
8. The child may continue with the prepared slips.

**NOTE:**

These equations suggest to the child the associative law of addition.

**Aim**

**Direct**

Development of order, concentration, coordination, independence and exactness.

Use the associative law of addition with multiple addends.

**Age**

6 years.

**Presentation 2 (Parenthesis)**

1. Select a prepared slip with an addition equation of three addends: 3 +5+7=.
2. Select the appropriate bead bars to represent this equation and place them horizontally with the + and = labels placed appropriately between the bead bars.
3. Introduce the parentheses and place them around the first two addends.
4. Add the beads within the parentheses first and place an eight-bar below the three-bar and five-bar in the equation.
5. Place a seven bead-bar below the seven bead-bar in the first equation.
6. Sum the second equation and place the sum beads vertically after the = sign.
7. Record this equation in the child's math notebook:

   (3 + 5) + 7 = 15

   8 + 7 = 15
8. The child may continue with the prepared slips.

**NOTE:**

In this presentation the child is introduced to signs (parentheses) which direct the child to apply the law of association to two particular addends.

## Aim

### Direct

Development of order, concentration, coordination, independence and exactness.

To introduce the parenthesis sign.

## Age

6 years.

## Presentation 3 (Parentheses in All Positions)

1. Select a prepared slip with an addition equation of four addends: 5 + (7 + 2) + 6 =.
2. Select the appropriate bead bars to represent this equation and place them horizontally with the + and = labels and parentheses placed appropriately between the bead bars.
3. Add the beads within the parentheses, first placing a nine-bar below the seven-bar and the two-bar.
4. Place a five bead-bar and a six bead-bar below the same bead bars in the first equation.
5. Sum the addends and place the sum beads vertically after the = sign.
6. Record this equation in the child's math notebook:

   5 + (7 + 2) + 6 =

   5 +  9  + 6 = 20
7. The child may continue with the prepared slips which place the parentheses in the possible positions:

   (5 + 7) + 2 + 6 =

   5 + (7 + 2) + 6 =

   5 + 7 + (2 + 6) =

**Aim**

**Direct**

To explore the use of the parentheses.

**Age**

6 years.

**Presentation 4 (Two Parentheses)**

1. Select a prepared slip with an addition equation of six addends and two sets of parentheses: (5 + 7) + (2 + 6) + 4 + 1 =.
2. Select the appropriate bead bars to represent this equation and place them horizontally with the + and = labels and parentheses placed appropriately between the bead bars.
3. Add the beads within the parentheses first: place a ten bead-bar and a two bead-bar below the five and seven; place an eight bear-bar below the two and six.
4. Place a four bead-bar and a one bead-bar below the same bead bars in the first equation.
5. Sum the addends and place the sum beads vertically after the = sign.
6. Record this equation in the child's math notebook:

   (5 + 7) + (2 + 6) + 4 + 1 =

   12 + 8 + 4 + 1 = 25
7. The child may continue with the prepared slips and two sets of parentheses.

**Aim**

**Direct**

Development of order, concentration, coordination, independence and exactness.

To further explore the use of parentheses.

**Age**

6 years.

**Presentation 5**

1. Select a prepared slip with an addition equation of three addends: 6 + 5 + 9 =.
2. Select the appropriate bead bars to represent this equation and place them horizontally with the + and = labels placed appropriately between the bead bars.
3. Sum the addends and place the sum beads vertically after the = sign. Allow this equation to lay on the work area.
4. Select a prepared slip with an addition equation of six addends and three sets of parentheses which is the "breakdown" of the previous equation:

   (4 + 2) + (1 + 4) + (3 + 6) =
5. Select the appropriate bead bars to represent this equation and place them horizontally with the + and = labels placed appropriately between the bead bars.
6. Add the beads within the parentheses, placing a six bar below the 4 + 2, a five-bar below the 1 + 4, and a nine-bar below the 3 + 6.
7. Sum the addends and place the sum beads vertically after the = sign.

8. Record the equations in the child's math notebook:

$$6 \quad + \quad 5 \quad + \quad 9 \quad = 20$$

$$(4 + 2) + (1 + 4) + (3 + 6) = 20$$

$$6 \quad + \quad 5 \quad + \quad 9 \quad = 20$$

9. The child may continue with the set of prepared slips.

**NOTE:**

The first equation is identical to the final equation.

**Age**

5 years.

# BASIC FORMATS FOR ADDITION

**Material**

Prepared slips with all possible equations for each of the following formats:

| | | | | | |
|---|---|---|---|---|---|
| ___ | + | 7 | = | 10 | (Calculate for the addend.) |
| 10 | = | ___ | + | 7 | (Calculate for the addend.) |
| 3 | + | ___ | = | 10 | (Calculate for the addend.) |
| 10 | = | 3 | + | ___ | (Calculate for the addend.) |
| ___ | = | 3 | + | 7 | (Calculate for the sum.) |
| 10 | = | ___ | + | ___ | (Calculate for the addend and the addend.) |

**Presentation**

1. Select a prepared equation slip of the first format and record it in the math notebook.
2. Determine the factor that is not given and record it in red.
3. Upon completion of the prepared equation slips, the Addition Control Chart 1 is used to compare to the answers in the math notebook for accuracy.
4. Continue in this manner for each of the equation formats.
5. The child may generate his own equations.

**Aim**

**Direct**

Development of order, concentration, coordination, independence and exactness.

Place the addition facts in their various formats.

**Age**

6 to 7 years.

# ADDITION STORY PROBLEMS

**Material**

General story problems which ask for one of each of the factors in an addition equation:

Addend:

> Susie has some oranges. Dan has five oranges. Altogether they have eight oranges. How many does Susie have?

Addend:

> Jim brought five bottles to school. Mary brought some bottles, also. Altogether they have nine bottles. How many bottles did Mary bring?

Surn:

> Alice brought five worms to fish with. George brought three worms. How many worms did they have altogether?

**Presentation**

1. Select a story problem and read it.
2. Ask:
   - A. How many does the first have? What is the addend?
   - B. How many does the second have? What is the addend?
   - C. How many altogether? What is the sum?
3. As the questions above are asked, the following blanks may be filled in:

   ___ + ___ = ___
4. The unfilled blank is the factor to be determined in the equation.
5. Record the equation in the math notebook.
6. Upon completion of the story problems, the Addition Control Chart 1 is used to compare the answers in the math notebook for accuracy.

**Aim**

**Direct**

Development of order, concentration, coordination, independence and exactness.

To place the addition facts into the context of practical life experience.

**Age**

6 to 7 years.

# INTRODUCTION TO THE SUBTRACTION STRIP BOARD

## Material

A board divided into eighteen 2-cm. squares across and twelve 2-cm. squares down. The vertical columns are numbered one through eighteen. One through nine are in blue, and ten through eighteen in red. A blue vertical line after the nine column divides the board.

Eighteen strips of natural colored wood of 2 cm., 4 cm., 6 cm., 8 cm., 10 cm., 12 cm., 14 cm., 16 cm., 18 cm., 20 cm., 22 cm., 24 cm., 26 cm., 28 cm., 30 cm., 32 cm., 34 cm., and 36 cm. long.

Nine strips of blue wood from 2 cm to 18 cm. long with a red numeral at the right end of each.

Nine strips of red wood from 2 cm. to 18 cm. long with a line in blue every 2 cm.

## Presentation

1. Place the board in the middle of the table and examine it with the child, noting the similarities to the addition strip board.
2. The blue vertical line comes after the nine. With this material, the answer (remainder) cannot be more than nine. It is the largest remainder we can have when we subtract 2 digits.
3. Lay out the natural wood strips and grade them from longest to shortest. These strips are used to establish the minuend.
4. Lay out the blue strips and grade them from longest to shortest. The red strips are not brought out.
5. Select an equation: 16 - 8 =.
6. Select the appropriate tan strip to establish the 16 as a minuend at the top of

the subtraction strip board. Lay the short strip over the numeral 17 and 18.

7. Select the blue 8-strip to represent the subtrahend and place it next to the tan strip, covering the numerals 16, 15, 14, 13, 12, 11, 10, and 9.
8. The remainder appears on the subtraction strip board to the left of the blue strip: 8.
9. The child may continue with the set of prepared slips.

**NOTE:**

The natural wood strip and the blue strip are both placed on the numerals along the top line of the subtraction strip board.

**Aim**

**Direct**

To begin the work of memorization of the subtraction facts.

**Age**

6 years.

# SUBTRACTION STRIP BOARD

**Material**

The subtraction strip board

Tan and blue strips

Booklet of printed equations

First subtraction control chart

**Presentation 1 (Booklet)**

1. Examine the booklet noting there is a page for each minuend from 18 to 1. The number of equations varies from page to page depending on the minuend. Remember, the remainder may not be more than 9. Therefore, the first page has only one equation:

   18- 9 =

   The next equation of:

   18 - 8 =

   would produce a remainder of 10, and with this material there can be no remainder more than 9.

   The second page has two equations:

   17 - 9 =

   17 - 8 =

   And so on for eighteen pages.

2. Begin with the first page of the booklet:

   18 - 9 =

3. Select the blue 9-strip for the subtrahend and place it over the numerals on the subtraction strip board, covering the numerals, 18, 17, 16, 15, 14, 13, 12, 11, and 10.

4. The remainder appears on the subtraction strip board to the left of the blue strip: 9.
5. Record the remainder.
6. Continue in this manner to the next page with:

   17 - 9 =

In this equation the natural wood strips are required to establish the minuend.

7. Continue with each page.
8. Upon completion of the booklet, the equations are checked by referring to Subtraction Control Chart 1.
9. Examine the Subtraction Control Chart 1:
   - A. All the remainders are in red.
   - B. On the left half of the chart on the first row all the remainders are 0. On the left half of the chart on the second row all the remainders are 1, and so on until 8.
   - C. On the right half of the chart on the first row the remainders increase by one from 0 to 9.

   In each row the minuend increases by one.

   Equal remainders are on the diagonal.
10. Compare each equation in the booklet to the Subtraction Control Chart 1 to determine correctness of the remainder recorded.

**Aim**

**Direct**

Development of order, concentration, coordination, independence and exactness.

To memorize the subtraction facts.

**Age**

6 to 7 years.

**Presentation 2 (Prepared Equation Slips)**

1. The child selects an equation and records it in his math notebook.
2. The natural wood strip is selected to establish the minuend at the top of the subtraction strip board.
3. The blue strip is selected for the subtrahend and placed to the left of the tan strip.
4. The remainder is the numeral to the left of the blue strip.
5. Record the remainder in the math notebook.
6. Replace the natural wood and blue strips.
7. Continue with the next prepared equation slip, recording the work as it progresses.
8. Upon completion of the prepared equation slips, compare the answers in the math notebook to the Subtraction Control Chart 1.

**Aim**

**Direct**

Development of order, concentration, coordination, independence and exactness.

To memorize the subtraction facts.

**Age**

6 years.

**Presentation 3 (Analysis of Number: Units)**

1. Say, "Let's see how we can analyze 9 on this subtraction strip board."
2. Set up the materials as before, but this time lay out the red strips in graded order from longest to shortest.
3. Select the natural wood strip to establish the minuend: 9.
4. Record the equation:

   9 - 1 = ___.

5. Place the blue l-strip on the first row of squares under the numeral 9.
6. These are 8 squares that remain to the left of the blue strip. Select the red 8-strip and place it to the left of the blue strip.
7. Record the remainder: 8.
8. Leave the strips on the board.
9. Record the equation:

   9 - 2 = ___.

10. Place the blue 2-strip on the second row of squares under the blue 1-strip.
11. There are 7 squares that remain to the left of the blue 2-strip. Select the red 7-strip and place it to the left of the blue 2-strip.
12. Record the remainder: 7.
13. Leave the strips on the board.
14. Record the equation:

    9 - 3 = ___.

15. Continue as before and progressively analyze 9:

    9 - 4 = ___

    9 - 5 = ___

    9 - 6 = ___

9 - 7 = ___

9 - 8 = ___

9 - 9 = ___

16. Upon completion of the analysis of 9, the Subtraction Control Chart 1 is used to compare to the answers in the math notebook for accuracy.
17. Continue on with the analysis of 8, 7, 6, 5, 4, 3, 2, and 1.

**NOTE:**

As the subtrahend increases, the remainder decreases.

**Aim**

**Direct**

To memorize the subtraction facts.

**Age**

6 to 7 years.

**Presentation 4 (Analysis of Number: Teens)**

1. Say, "Let's see how we can analyze 15 on the subtraction strip board."
2. Set up the natural wood strips, the blue strips, and the red strips.
3. Select the natural wood strips to establish the minuend: 15.
4. Record the equation:

   15 - 1 = ___

5. Place the blue l-strip on the first row of squares under the numeral 15.

   Note that the remainder is more than nine. Cross the equation off in the math notebook.

6. Record the equation:

   14 - 2 = ___

7. Place the blue 2-strip on the first row of squares under the numeral 15. Note that the remainder is more than nine, so this equation cannot be done. Cross the equation off in the math notebook.

8. Continue on with the analysis:

   15 - 3 = ___

   15 - 4 = ___

   15 - 5 = ___

   These equations are not possible within the confines of not having a remainder more than 9. Therefore, they are crossed off in the math notebook.

9. Then proceed as in Presentation 3 for the rest of the analysis of 15:

   15 - 8 = ___

   15 - 9 = ___

10. Upon completion of the analysis of 15, the Subtraction Control Chart 1 is used to compare to the answers in the math notebook.

11. Continue on with the analysis of 18, 17, 16, 15, 14, 13, 12, 11, and 10.

**Aim**

**Direct**

Development of order, concentration, coordination, independence and exactness.

To memorize the subtraction facts.

**Age**

6 to 7 years.

**Presentation 5 (Zero)**

1. Say, "Let's see how we can analyze 9 on the subtraction strip board, but this time we shall do something different."
2. Select the natural wood strip to establish the minuend: 9.
3. Record the equation:

   9 - 0 =

4. No blue strip is required because the blue strips are subtrahends and the subtrahend is zero.
5. Select the red 9-strip and place it on the first row of squares.
6. Record the remainder: 9.

7. Leave the strips on the board.
8. Record the equation:

   9 - 1 =

9. Proceed as in Presentation #3, continuing through the complete analysis of 9:

   9 - 2 = ___

   9 - 3 = ___

   9 - 4 = ___

   9 - 5 = ___

   9 - 6 = ___

   9 - 7 = ___

   9 - 8 = ___

   9 - 9 = ___

10. Upon completion of the analysis of 9 with zero, the Subtraction Control Chart 1 is used to compare to the answers in the math notebook for accuracy.
11. Continue on with the analysis of 18 through 1 with zero.

**Aim**

**Direct**

Development of order, concentration, coordination, independence and exactness.

To memorize the addition facts.

**Age**

6 to 7 years.

**Presentation 6 (Subtraction Chart 2)**

1. Examine Subtraction Chart 2. The minuends are at the top of the chart; the subtrahends are on the diagonal along the left.
2. Select a prepared equation slip and record it in the math notebook.
3. With the index finger of the right hand, find the minuend along the top of the Subtraction Chart 2.
4. With the index finger of the left hand, find the subtrahend on the left diagonal.
5. Move the right index finger down and the left index finger across. Where they meet is the remainder.
6. Record the remainder.
7. Select another prepared equation slip and proceed as before.
8. Upon completion of the prepared equation slips, the Subtraction Control Chart 1 is used to compare to the answers in the math notebook for accuracy.

**Aim**

**Direct**

Development of order, concentration, coordination, independence

and exactness.

To memorize the subtraction facts.

**Age**

6 to 7 years.

**Presentation 7 ("Bingo Game #1")**

**Material**

Subtraction Chart 3 (Blank Chart)

Stamps with remainders

Prepared equation slips

Control Charts 1 and Chart 2

Math notebook

**Presentation**

1. Lay all the stamps face up on the table.
2. Select an equation and record it in the math notebook.
3. From the stamps, select the correct remainder and place it at the top of the Subtraction Chart 3.
4. With the index finger of the right hand, find the minuend along the top of Subtraction Chart 3.
5. With the index finger of the left hand, find the subtrahend along the left diagonal.
6. Move the right index finger down and the left index finger across. Where they meet is the blank square where the remainder stamp should be placed.
7. Record the remainder in the math notebook.
8. Select another prepared equation slip and proceed as before.

9. Upon completion of Subtraction Chart 3, compare it to Subtraction Chart 2. They should be identical. Compare the equations in the math notebook to Subtraction Chart 1 for accuracy.

**Aim**

**Direct**

To memorize subtraction facts.

**Age**

6 to 7 years.

**Presentation 8 ("Bingo Game #2")**

1. Lay all the stamps face up on the table.
2. Select a remainder (stamp) at random and place it at the top of Subtraction Chart 3. Record it in the math notebook.
3. The child makes up an equation to satisfy the remainder and enters the complete equation in the math notebook.
4. With the index finger of the right hand, find the minuend along the top of the Subtraction Chart 3.
5. With the index finger of the left hand, find the subtrahend along the left diagonal.
6. Move the right index finger down and the left index finger across. Where they meet is the blank square where the remainder stamp should be placed.
7. Select another remainder (stamp) and proceed as before.
8. Upon completion of Subtraction Chart 3, compare it to Subtraction Chart 2. They should be identical. Compare the equations in the math notebook to Subtraction Chart 1 for accuracy.

**Aim**

**Direct**

To memorize the subtraction facts.

**Age**

6 to 7 years.

**Presentation 9 ("Bingo Game #3")**

1. The child lays all the stamps face up in stacks of the same denomination from 1 through 9.

   NOTE: There are nine of each numeral 1 through 9.
2. Select a remainder stack (stamps).
3. Make up the nine possible subtraction equations to satisfy the remainder. Each equation is entered in the math notebook, and with the index fingers, the position of the remainder on the Subtraction Chart 3 is located.
4. Upon completion of the stacks of remainders (stamps), compare the filled-in Chart 3 (blank chart) with Chart 2. They should be identical. Compare the equations in the math notebook to Subtraction Control Chart 1 for accuracy.

**Aim**

**Direct**

To memorize the subtraction facts.

**Age**

6 to 7 years.

# ORAL GAMES FOR THE MEMORIZATION OF SUBTRACTION

**Material**

Subtraction equation slips

Remainder stamps

**Presentation 1**

1. The child selects a subtraction equation slip and reads it.
2. Another child says the remainder.
3. The first child determines the accuracy of the response from memory or from Control Chart 1.
4. Proceed with the other equation slips in this manner.

**Aim**

**Direct**

Development of order, concentration, coordination, independence and exactness.

To memorize the subtraction facts.

**Age**

6 to 7 years.

**Presentation 2**

1. The child selects a remainder (stamp) and reads it.
2. Another child says the possible equations to satisfy this remainder.
3. The first child determines the accuracy of the responses from memory or from

Control Chart 1.

4. Proceed with the other remainder stamps in this manner.

## Aim

### Direct

Development of order, concentration, coordination, independence and exactness.

To memorize the subtraction facts.

## Age

6 to 7 years.

# BASIC FORMATS FOR SUBTRACTION

**Material**

Prepared slips with all possible equations for each of the following formats:

| | | | | | |
|---|---|---|---|---|---|
| ___ | - | 7 | = | 3 | (Calculate for the minuend.) |
| 3 | = | ___ | - | 7 | (Calculate for the minuend.) |
| 10 | - | ___ | = | 3 | (Calculate for the subtrahend.) |
| 3 | = | 10 | - | ___ | (Calculate for the subtrahend.) |
| ___ | = | 10 | - | 7 | (Calculate for the remainder.) |
| 3 | = | ___ | - | ___ | (Calculate for the minuend and |
| 3, 7, 10 | | | | | (Write as many equations as you can using these numbers.) |

**Presentation**

1. Select a prepared equation slip of the first format and record it in the math notebook.
2. Determine the factor that is not given and record it in red.
3. Upon completion of the prepared equation slips, the Subtraction Control Chart 1 is used to compare to the answers in the math notebook for accuracy.
4. Continue in this manner for each of the equation formats.
5. The child may generate his own equations.

**Aim**

**Direct**

Development of order, concentration, coordination, independence and exactness.

Place the subtraction facts in their various formats.

**Age**

6 to 7 years.

# SUBTRACTION STORY PROBLEMS

**Material**

Generate story problems which ask for one of each of the factors in a subtraction equation:

Minuend:

Mary gave four of her toy trucks to Ann. She then had three left. How many trucks did she have when she began?

Subtrahend:

John had eight balloons. He gave some to Elsa. He then had three remaining. How many did he give to Elsa?

Remainder:

Eddie had nine cookies. He gave seven to Bob. How many did he have left?

Presentation

1. Select a story problem and read it.
2. Ask:
   - A) How many are there altogether? What is the minuend?
   - B) How many are taken away? What is the subtrahend?
   - C) How many are left? What is the remainder?
3. As the questions above are asked, the following blanks may be filled in:

   ___ - ___ = ___.
4. The unfilled blank is the factor to be determined in the equation.
5. Record the equation in the math notebook.
6. Upon completion of the story problems, the Subtraction Control Chart 1 is used to compare to the answers in the math notebook for accuracy.

## Aim

### Direct

Development of order, concentration, coordination, independence and exactness.

To place the subtraction facts into the context of practical life experience.

## Age

6 to 7 years.

# SUBTRACTION SNAKE GAME

**Material**

A box containing colored bead bars for the numerals one to nine. There should be eight to ten bars for each number.

One - red

Two - green

Three - pink

Four - yellow

Five - blue

Six - lavender

Seven- white

Eight - brown

Nine - dark blue

A box with a set of black and white bead bars. All the bead bars through five are black. From six through nine the bead bars are black for the first five beads and white for successive beads.

A box of ten-bead bars

A notched strip of cardboard

A box of golden bead bars

A box with several sets of grey bead bars

**Presentation**

1. Remove the black and white bead bars and place them to the upper right of the work area in graded order.
2. Select a small arrangement of colored bead bars and lay them "snakewise" across the work area.

3. Select the grey bead bars and explain to the child that these bead bars are for subtraction.
4. Space the grey bead bars between the colored beads.
5. Begin to add the colored bead bars as in the addition snake game.
6. When arriving at a grey bead bar, stop and say, "This means I must subtract this amount." Determine the value of the grey beads.
7. Then move down the grey bar and the colored bar preceding it. The bars should be subtracted without counting. State the subtraction fact orally. From the child's knowledge of the subtraction facts, the difference should be stated. If the colored bar is less than the grey bar, move down the next preceding colored or golden bar until there is a sufficient amount to form a subtraction problem. Replace the bars which formed the subtraction problem with a black and white bar equal to the difference.
8. The golden beads that are removed are placed back in their box. The colored bead bars and the grey bead bars are used for verification and are placed to the side.
9. Continue until the snake has been changed to golden beads and all the grey bead bars have been removed.
10. Verification: "Let's see if I have counted correctly." Line up the ten bead bars to the left. Line up the colored bead bars to the right, group colors together. Line up the grey bead bars in graded fashion. Begin by matching the grey bead bars to the colored bead bars. When that is complete, match the colored bead bars to the ten-bead bars. If this comes out even, then the counting of the subtraction snake is accurate.
11. Return the bead bars to their boxes and to the tray.

**NOTE:**

As the subtraction snake is counted it becomes shorter and shorter. Always begin the snake with enough colored bead bars so that negative numbers are not encountered.

**Aim**

**Direct**

Development of order, concentration, coordination, independence and exactness.

To memorize the subtraction facts.

**Age**

6 to 7 years.

# INTRODUCTION TO THE MULTIPLICATION BEAD BOARD

**Material**

Multiplication Bead Board

A box of 100 red beads

Small white numeral cards 1 to 10

A red disc

A booklet of equations: multiplication tables 1 to 10

Prepared equation slips

Five charts

Math notebook

**Presentation**

1. Place the board in the middle of the table and examine it with the child.
2. The numeral on the left side that fits into the slot is called the multiplicand. It tells how many are in a set.
3. The numerals from 1 to 10 along the top of the board are called the multiplier. They tell how many times to make the set.
4. The red disc rests in the slot in the upper left corner. It is placed above the multiplier when working an equation as a reminder or indicator of which numeral is the multiplier.
5. Select a prepared equation slip:

   4 x 6 = ___.
6. Record the equation in the math notebook.
7. Place the numeral six in the slot on the left to stand for the multiplicand.
8. Place the red disc above the numeral four to stand for the multiplier.

9. The equation reads: 4 sets of 6.
10. Move the red disc above the one and count out six beads under the numeral one.
11. Move the red disc above the numeral two and count out six beads under the numeral two.
12. Move the red disc above the numeral three and count out six beads under the numeral three.
13. Move the red disc above the numeral four and count out six beads under the numeral four.
14. Pointing to the multiplier and multiplicand, say, "4 sets of 6," and count up the number of beads on the board to complete the equation.
15. Record the product in the math notebook.
16. The child may continue with equations to practice on the multiplication board.

**NOTE:**

The concept of multiplication has already been introduced with the golden bead material where the same quantity is added together several times. With the multiplication bead board, the multiplication facts are memorized for ease of calculation.

**Aim**

**Direct**

Development of order, concentration, coordination, independence and exactness.

To memorize the multiplication facts.

**Age**

6 to 7 years.

# MULTIPLICATION BEAD BOARD

**Material**

Complete set of materials for the multiplication bead board

Booklet of equations

Multiplication Control Chart 1

**Presentation 1 (Booklet of Prepared Equations)**

1. Examine the booklet, noting that there is a new page for each multiplicand, and each multiplicand is multiplied 1 through 10.
2. Turn to page 3.
3. Establish the multiplicand 3 by placing the numeral 3 in the slot to the left.
4. Read the first equation: 1 x 3 = ___. Place the disc over the 1 on the board, and place 3 beads under it.
5. Record the answer: 1 x 3 = 3.
6. Read the next equation: 2 x 3 = ___.
7. Move the disc to the 2 and place a second set of 3 beads under it, so there are now 6 beads on the board.
8. Record the answer: 2 x 3 = 6.
9. Continue in the same way to 10 x 3 = 30.
10. Upon completion of the booklet or a page of the booklet, the equations are checked for accuracy by referring to Multiplication Control Chart 1.

**NOTE:**

The maximum number of products in any base is (b - 1)2. In base 10: (10 - 1)2 = 81. Therefore,the maximum number of products the child should have to memorize should be 81. However, Dr. Montessori went on to the table of ten to emphasize how easy

the base ten system is. The only difference between the table of one and the table of ten is the recurrence of zero.

## Aim

### Direct

Development of order, concentration, coordination, independence and exactness.

To memorize the multiplication facts.

## Age

6 to 7 years.

# MULTIPLICATION BEAD BOARD: PREPARED EQUATION SLIPS

**Material**

Complete set of materials for the multiplication bead board

Multiplication prepared equation slips

Multiplication Control Chart 1

**Presentation**

1. The child lays out all the numerals 1 to 10 representing the multiplicand.
2. Select an equation: 4 x 7 = ___.
3. Record the equation in the math notebook.
4. Establish the multiplicand by inserting the 7 card in the slot.
5. Establish the multiplier by placing the disc above the 4 on the board.
6. Move the disc to the 1, make 1 x 7 beads; move the disc to the 2, make 2 x 7 beads, etc., to 4 x 7.
7. Count the beads and record the answer: 4 x 7 = 28.
8. Clear the board, removing beads, disc, and card.
9. Choose another equation.
10. Proceed to do this equation as before.
11. Upon completion of the prepared equation slips, the equations are checked for accuracy by referring to Multiplication Control Chart 1.

**NOTE:**

Working with the beads and board, the child forms geometric figures: when the two factors are the same, a square is formed; when the two factors are different, a rectangle is formed.

12. Examine Multiplication Control Chart 1. The same products run diagonally from bottom to top, left to right.

13. Begin at the upper left-hand equation: 1 x 1 = 1. There is no duplicate for that equation.

14. Move one equation to the right: 2 x 1 = 2.

    Move one equation down: 1 x 2 = 2.

    Because these are duplicates, eliminate 2 x 1 = 2. Place a blank, rectangular paper over 2 x 1 = 2.

15. Continue in this manner until all duplicate equations are eliminated and the upper right equations are eliminated from the diagonal.

16. Compare Multiplication Control Chart 2 with the Multiplication Chart 1 that has been covered over. They are identical with 55 equations.

**NOTE:**

The squares of the number appear along the diagonal of the Multiplication Chart 2. An alternative activity for the elimination of duplicate equations is having the child copy Chart 1 and erase duplicates, thereby, generating Chart 2.

### Aim

#### Direct

Development of order, concentration, coordination, independence and exactness.

To memorize the multiplication facts.

### Age

6 to 7 years.

# MULTIPLICATION BEAD BOARD: COMBINATIONS OF NUMBER

**Material**

Complete set of materials for the multiplication bead board

Multiplication Control Chart 1

Product stamps

**Presentation**

1. Select a product stamp.
2. Record the product in the math notebook.
3. Count out in beads the product indicated.
4. Generate equations that satisfy the product and lay them out on the multiplication bead board.
5. Upon completion, the equations are checked for accuracy by referring to Multiplication Control Chart 1.

**Aim**

**Direct**

Development of order, concentration, coordination, independence and exactness.

To memorize the multiplication facts.

**Age**

6 to 7 years.

# MULTIPLICATION CHART 3

**Material**

Prepared equation slips

Multiplication Charts 1 and 3

Math notebook

**Presentation**

1. Examine Multiplication Chart 3. The multiplicand is along the top; the multiplier is down the left side; the products appear in the matrix of the chart.
2. Select a prepared equation slip.
3. Record the equation in the math notebook.
4. With the right index finger, find the multiplicand at the top of the chart.
5. With the left index finger, find the multiplier along the left side of the chart.
6. Bring the right index finger down and the left index finger across. The square in which the fingers meet is the product.
7. Record the product in the math notebook.
8. Upon completion, the equations are checked for accuracy by referring to Multiplication Control Chart 1 or 2.

**Aim**

**Direct**

Development of order, concentration, coordination, independence and exactness.

To memorize the multiplication facts.

**Age**

6 to 7 years.

# MULTIPLICATION CHART 4

**Material**

Prepared equation slips

Multiplication Chart 1

Multiplication Chart 4

Math notebook

**Presentation**

1. Examine Multiplication Chart 4. The top row has been eliminated and both multiplicand and multiplier are represented along the left side. The products that represent the commutative law of multiplication have been eliminated so the upper left half of the chart is eliminated.
2. Select a prepared equation slip.
3. Record the equation in the math notebook.
4. With the right index finger, locate the smaller of the two numerals in the equation along the left side of the chart.
5. With the left index finger, locate the larger numeral in the equation along the left side of the chart.
6. Bring the right finger as far across the chart as possible.
7. Move the left finger to directly below the right finger. Move the right finger down to meet the left finger. This is the product.
8. Record the product in the math notebook.
9. Upon completion, the equations are checked for accuracy by referring to Multiplication Chart 1.

## Aim

### Direct

Development of order, concentration, coordination, independence and exactness.

To memorize the multiplication facts.

## Age

6 to 7 years.

# MULTIPLICATION CHART 5 ("BINGO CHART" or "BLANK CHART")

**Material**

Prepared equation slips

Product stamps

Multiplication Control Chart 1

Multiplication Charts 3 and 5

Math notebook

**Presentation 1**

1. Examine Multiplication Chart 5. It is a blank chart with a matrix containing 100 squares with numerals from 1 to 10 across the top and along the left side.
2. Lay all the stamps face up on the table.
3. Select an equation and record it in the math notebook.
4. From the stamps, select the correct product and place it at the top of the Multiplication Chart 5.
5. With the index finger of the right hand, find the multiplicand along the top of the chart.
6. With the index finger of the left hand, find the multiplier along the left hand side of the chart.
7. Move the right index finger down and the left index finger across. Where they meet is the blank square where the product stamp should be placed.
8. Record the product in the math notebook.
9. Select another prepared equation slip and proceed as before.
10. Upon completion of Multiplication Chart 5, compare it to Multiplication Chart 3. They should be identical. Compare the equations in the math notebook to

Multiplication Control Chart 1 for accuracy.

## Aim

### Direct

Development of order, concentration, coordination, independence and exactness.

To memorize the multiplication facts.

## Age

6 to 7 years.

## Presentation 2

1. Lay all the stamps face up on the table.
2. Select a product (stamp) at random and place it at the top of Multiplication Chart 5. Record it in the math notebook.
3. The child makes up an equation to satisfy the product and enters the complete equation in the math notebook.
4. With the index finger of the right hand, find the multiplicand along the top of the chart.
5. With the index finger of the left hand, find the multiplier along the left side of the chart.
6. Move the right index finger down and the left index finger across. Where they meet is the blank square where the product stamp should be placed.
7. Select another product stamp and proceed as before.
8. Upon completion of Multiplication Chart 5, compare it to Multiplication Chart 3. They should be identical. Compare the equations in the math notebook to Multiplication Control Chart 1 for accuracy.

**Aim**

**Direct**

Development of order, concentration, coordination, independence and exactness.

To memorize the multiplication facts.

**Age**

6 to 7 years.

**Presentation 3**

1. The child lays all the stamps face up in stacks of the same denomination from low numerals to high numerals.
2. Select a product (stamps) stack.
3. Make up all the equations that would satisfy this product. The number of stamps in the stack indicates the number of possible equations.
4. Enter each equation in the math notebook.
5. With the index fingers, locate the position of the product on the Multiplication Chart 5.
6. Upon completion of the stacks of products (stamps), compare the filled-in Chart 5 with Chart 3. They should be identical. Compare the equations in the math notebook to Multiplication Control Chart 1 for accuracy.

**Aim**

**Direct**

To memorize the multiplication facts.

**Age**

6 to 7 years.

# ORAL GAMES FOR THE MEMORIZATION OF MULTIPLICATION

**Material**

Multiplication equation slips

Product stamps

**Presentation 1**

1. The child selects a multiplication equation slip and reads it.
2. Another child says the product.
3. The first child determines the accuracy of the response from memory or from Control Chart 1.
4. Proceed with the other equation slips in this manner.

**Aim**

**Direct**

Development of order, concentration, coordination, independence and exactness.

To memorize the multiplication facts.

**Age**

6 to 8 years.

**Presentation 2**

1. The child selects a product (stamp) and reads it.
2. Another child says the possible equations to satisfy this product.
3. The first child determines the accuracy of the response from memory or from

Control Chart 1.

4. Proceed with the other product stamps in this manner.

## Aim

### Direct

Development of order, concentration, coordination, independence and exactness.

To memorize the multiplication facts.

## Age

6 to 8 years.

# BASIC FORMATS FOR MULTIPLICATION

**Material**

Prepared slips with all possible equations for each of the following formats:

| | | | | | |
|---|---|---|---|---|---|
| ___ | x | 7 | = | 21 | (Calculate for the multiplier.) |
| 21 | = | ___ | x | 7 | (Calculate for the multiplier.) |
| 3 | x | ___ | = | 21 | (Calculate for the multiplicand.) |
| 21 | = | 3 | x | ___ | (Calculate for the multiplicand.) |
| ___ | = | 3 | x | 7 | (Calculate for the product.) |
| 21 | = | ___ | x | ___ | (Calculate for the multiplicand and mutliplier.) |

**Presentation**

1. Select a prepared equation slip of the first format and record it in the math notebook.
2. Determine the factor that is not given and record it in red.
3. Upon completion of the prepared equation slips, the Multiplication Control Chart 1 is used to compare the answer in the math notebook for accuracy.
4. Continue in this manner for each of the equation formats.
5. The child may generate his own equations.

**Aim**

**Direct**

Place the multiplication facts in their various formats.

**Age**

6 to 8 years.

# MULTIPLICATION STORY PROBLEMS

**Material**

Generate story problems which ask for one of each of the factos in a multiplication equation:

Multiplicand:

There are 3 eggs in each basket. All together there are 24 eggs. How many baskets are there?

Multiplier:

Six boys are bringing cookies to the class party so that each child in the class will have a cookie. There are forty-two children in the class. How many cookies will each boy bring? Product:

Four children are building a hideout. Each child brings seven boxes. How many boxes will the children have altogether for their hideout?

**Presentation**

1. Select a story problem and read it.
2. Ask:
   - A) How many in each set? What is the multiplicand?
   - B) How many sets? What is the multiplier?
   - C) How many are there altogether? What is the product?
3. As the questions above are asked, the following blanks may be filled in:

   ___ x ___ = ___
4. The unfilled blank is the factor to be determined in the equation.
5. Record the equation in the math notebook.
6. Upon completion of the story problems, the Multiplication Control Chart 1 is used to compare to the answers in the math notebook for accuracy.

**Aim**

**Direct**

Development of order, concentration, coordination, independence and exactness.

To place the multiplication facts into the context of practical life experience.

**Age**

6 to 7 years.

# SNAKE GAME FOR MEMORIZATION OF MULTIPLICATION

**Material**

A box containing colored bead bars for the numerals one to nine. There should be eight to ten bars for each number.

One - red

Two - green

Three - pink

Four - yellow

Five - blue

Six - lavender

Seven- white

Eight - brown

Nine - dark blue

A box with a set of black and white bead bars. All the bead bars through five are black. From six through nine the bead bars are black for the first five beads and white for successive beads. This allows for ease in identifying the quantity on the bead bar.

A box of ten bead bars

A notched strip of cardboard

A box of golden bead bars

**Presentation**

1. Remove the black and white bead bars and place them to the upper right of the work area in graded order.
2. Place the box of golden bead bars in the upper left of the work area.
3. Select an arrangement of colored bead bars and lay them "snake-wise" across

the work area. Use only three or four kinds of bead bars and use several of each.

4. Proceed as in the addition snake game to add the colored bead bars and change them to golden ten bars.
5. Verification: "Let's see if I have added correctly." Line up the ten bars and any remaining black and white beads to the left, and line up the colored bead bars to the right.
6. Bring down a set of colored bead bars that are the same. Multiply them to determine the total. Match an equal amount of golden bead bars to the colored bead bars. Exchange golden bead bars when necessary to obtain the correct quantity to match to the colored bead bars.
7. Continue as above for each set of colored bead bars that are the same.
8. When all the colored bead bars have been placed in one-to-one correspondence to the golden bead bars, verification is complete.

**Aim**

**Direct**

Development of order, concentration, coordination, independence and exactness.

To memorize the addition facts.

To make the relationship between addition and multiplication.

To show that multiplication is faster than repeated addition.

**Age**

5 to 6 years.

# CUBING CHAINS FOR THE MEMORIZATION OF MULTIPLICATION

**Material**

Wooden bead cabinet with top ledge to hold the cubes from left to right: ten cube, nine cube . . . one cube. Hooks extending down from the top ledge of the cabinet to hold the cubing chains from left to right: ten cubing chain, nine cubing chain . . . one cubing chain.

From the right side of the cabinet, ledges extend out in increasing lengths so that the tenth ledge forms the bottom border of the bead cabinet. The top ledge on the right contains the one square and one squaring chain. And each ledge contains the next number in sequence down to the bottom ledge, which contains the ten ten-squares and the ten squaring chain.

Labels organized by cubing chains.

**Presentation 1**

1. Bring a single bead, a three bead-bar, the three squaring chain, and the three cube.
2. Lay out the single bead and say, "This is a bead."
3. Lay out the three bead-bar and say, "This is a bead bar."
4. Lay out the three squaring chain and say, "This is a squaring chain."
5. Lay out the three bead-square and say, "This is a bead square."
6. Lay out the three cubing chain and say, "This is a cubing chain."
7. Lay out the three bead-cube and say, "This is a bead cube."
8. With the three-period lesson and discussion, repeat the nomenclature.

**Aim**

**Direct**

Development of order, concentration, coordination, independence and exactness.

To memorize the multiplication facts.

**Age**

5 to 6 years.

**Presentation 2**

1. Select a squaring chain. Usually the four, five, or six is best for illustration of the procedure.
2. Carry the chain with great care, one hand at each end of the chain.
3. Stretch the chain stretched out vertically.
4. Place the square at the top of the chain.
5. With care and dexterity, move one bar up at a time to form the square.
6. Superimpose the bead square over the square formed by the chain.
7. Replace the square at the top of the squared chain.
8. Stretch the chain out again.
9. Place the arrows in order to the right of the bead chain.
10. Count the beads, beginning at the top and placing the appropriate arrow by the bead at the end of each set.
11. After placing the final arrow, which is the square of the bead chain, place the bead square to the right of the arrow indicating the square.
12. Walking along the chain, count in sequence from the first set to the last set of bead bars.
13. Walk back counting backwards.

**Aim**

**Direct**

Development of order, concentration, coordination, independence and exactness.

To memorize the multiplication facts.

**Age**

5 to 6 years.

**Presentation 3**

1. Select a cubing chain. Usually the four, five, or six is best for illustration of the procedure.
2. Remove the chain from the cabinet by removing each loop from the top hook and placing it in the dominant hand. Grasp the other end with the subdominant hand if bead quantity permits, or drape the beads across the subdominant arm.
3. Carry the chain with great care.
4. Stretch the chain stretched out vertically.
5. Remove the squares of the chain, one at a time, from the bead cabinet and place them in a stack at the top of the chain.
6. Remove the cube of the chain and place it to the right of the stacked squares.
7. With great care and dexterity, move one bar of the chain up at a time to form a square.
8. Superimpose the bead square over the square formed by the chain.
9. Place the square to the right of the square formed by the bead bars.
10. Continue in this manner through all the bars of the chain, forming a square and superimposing a bead square, then placing the bead square to the right. In

this way, two long rectangles are formed: one of bead bars and one of bead squares.

11. "How many bead squares do we have?"
12. Beginning with the last bead square, stack the squares while counting. When the cube is formed with the squares say, "(x) squares equal a cube."
13. With care and dexterity, stretch the cubing chain out.
14. Place the arrows in order.
15. Count the beads, beginning at the top and placing the appropriate arrow at the bead. Continue until the end of the chain.
16. Walking along the chain, read the labels in sequence.
17. Walk back and read the labels backwards.

**Aim**

**Direct**

Development of order, concentration, coordination, independence and exactness.

To learn the numbers in linear counting as they are arranged in subsets.

**Indirect**

To memorize the multiplication facts through skip-counting.

**Age**

5 to 6 years.

**Presentation 4**

1. Both the squaring chain and cubing chain for a particular number are layed out.
2. The labels for both chains are placed into position.

3. The two chains are compared.

**Aim**

**Direct**

Development of order, concentration, coordination, independence and exactness.

To learn the numbers in linear counting as they are arranged in subsets.

To memorize the multiplication facts through skip-counting.

**Age**

5 to 6 years.

**Presentation 5 (Geometric Forms)**

1. Lay all ten bead-chains out.
2. Begin with the one chain, and say, "This is a point."
3. Lay out the two chain, form an angle, and say, "This is an angle."
4. Lay out the three chain, form a triangle, and say, "This is a triangle."
5. Lay out the four chain, form a square, and say, "This is a square."
6. Lay out the five chain, form a pentagon, and say, "This is a pentagon."
7. Lay out the six chain, form a hexagon, and say, "This is a hexagon."
8. Lay out the seven chain, form a heptagon, and say, "This is a heptagon."
9. Lay out the eight chain, form an octagon, and say, "This is an octagon."
10. Lay out the nine chain, form a nonagon, and say, "This is a nonagon."
11. Lay out the ten chain, form a decagon, and say, "This is a decagon."

**Aim**

**Direct**

Development of order, concentration, coordination, independence and exactness.

To learn the numbers in linear counting as they are arranged in subsets.

**Indirect**

To memorize the multiplication facts through skip-counting.

**Age**

5 to 6 years.

# BEAD BARS FOR THE MEMORIZATION OF MULTIPLICATION

**Material**

A box with ten compartments containing twenty bead bars in each compartment 1 - 10.

Labels with x and = signs.

**Presentation 1 (Multiplication with One Multiplicand)**

1. Select a bead bar, such as four.
2. Multiply four by all the numbers from 1 to 10.
3. Lay one bar of four horizontally, saying, "1 x 4 = 4," and place the product of four vertically below the horizontal four.
4. Lay out two bars of four horizontally to the right of the last beads, saying, "2 x 4 = 8," and place the product of eight vertically below the horizontal fours.
5. Continue in this manner to work all the equations up through 10 x 4 = ___.
6. Each equation may be recorded in the math notebook.
7. Labels may be made to place at the top of each bead equation:

   1 x 4 = 4 , 2 x 4 = 8 , etc.
8. Continue to make all the tables.
9. Upon completion of the tables, the Multiplication Control Chart 1 is used to compare to the answers in the math notebook for accuracy.

**Aim**

**Direct**

Development of order, concentration, coordination, independence and exactness.

To memorize the multiplication facts.

**Age**

6 to 7 years.

**Presentation 2 (Multiplication by Ten)**

1. Select ten of any of the bead bars and line them up horizontally, one under the other.
2. Place the product in golden bead bars vertically under the bead bars.
3. Record the complete equation in the math notebook.
4. Continue as above with each number 1 through 10.
5. Upon completion of these multiples of 10, the Multiplication Control Chart 1 is used to compare to the answers in the math notebook for accuracy.

**NOTE:**

The child may notice that multiplying by 10 is easiest, because a zero need only be added to the multiplicand. If this realization is not independently arrived at, point it out to the child after the child has had adequate experience with the material.

**Aim**

**Direct**

Development of order, concentration, coordination, independence and exactness.

To memorize the multiplication facts.

**Age**

6 to 7 years.

**Presentation 3 (All Factors for One Product)**

1. Select a product, such as 12.
2. Lay out twelve beads: one ten-bar and a colored two-bar laid vertically.
3. Begin with the one; see if it can be a factor in the equation. Can one be a factor? Yes, 12 sets of 1 is 12.
4. Lay out the two-bars. Two-bars taken six times equal twelve. The two-bars are laid to the right, horizontally.
5. Lay out the three-bars. Three-bars taken four times equal twelve. The three-bars are laid to the right, horizontally.
6. Continue with the rest of the bead bars and determine which bead bars are factors of the product 12.
7. Upon completion of the factors of a particular product, record them in the math notebook. The Multiplication Control Chart 1 is used to compare to the answers in the math notebook for accuracy.
8. Select another product, and beginning with the one bead bars determine the possible factors 1 through 10.

**NOTE:**

Some products have many factors. Some products have few factors. Some products have no factors within the range of 1 through 10.

**Aim**

**Direct**

Development of order, concentration, coordination, independence and exactness.

To memorize the multiplication facts.

**Age**

6 to 8 years.

**Presentation 4 (Small Multiplication)**

1. Select several bars of any number, but no more than ten.
2. Lay out one of the bars and record the equation and product.
3. Lay out a second bead bar and record the new equation and product formed.
4. Lay out the rest of the bead bars, one at a time, and record the equation in the math notebook.
5. The Multiplication Control Chart 1 is used to compare to the answers in the math notebook for accuracy.
6. Continue in this manner, selecting several bead bars and creating an abbreviated multiplication table.
7. Prepare tickets for the children to use. (5 sets of 6), (4 sets of 7), etc.

**Aim**

**Direct**

Development of order, concentration, coordination, independence and exactness.

To memorize the multiplication facts.

**Age**

6 to 8 years.

**Presentation 5 (Reverse Factors of the Same Product)**

1. Select several bead bars of any number. (5 x 7)
2. Lay out the bead bars and record the equation in the math notebook.

3. Select the reverse of the equation in bead bars (7 x 5) and lay them to the right of the first bead bars.
4. Record the second equation in the math notebook.
5. Continue in this manner with many selections and inversions.
6. Upon completion of the inverse equations, the Multiplication Control Chart 1 is used to compare to the answers in the math notebook for accuracy.

**Aim**

**Direct**

To memorize the multiplication facts.

**Age**

6 to 8 years.

**Presentation 6 (Square of Numbers)**

1. Form the square of each number, placing the necessary number of bars.
2. Record the equations in the math notebook.
3. Upon completion of the squares of the numbers, the Multiplication Control Chart 1 is used to compare to the answers in the math notebook for accuracy.
4. Ask the child what figure is formed. Place a square of the number on the square formed by the bead bars.

**NOTE:**

The square products appear along the diagonal in the Control Chart.

**Aim**

**Direct**

Development of order, concentration, coordination, independence and exactness.

To memorize the multiplication facts.

**Age**

6 to 8 years.

**Presentation 7 (Binomial by One Number)**

1. Select a binomial. (4 + 6)
2. Explain that you wish to multiply (4 + 6) by 4.
3. Place the multiplier and the x sign before the binomial and the = sign after the binomial: 4 x (4 + 6) = ___.
4. Record the equation in the math notebook.
5. Multiply the first term by four (4 x 4), select the bead bars, and lay them out.
6. Multiply the second term by four (4 x 6), select the bead bars, and lay them to the right of the first bead bars.
7. Place a + sign between them and an = sign after them.
8. Record the second part of the equation: (4 x 4) + (4 x 6) = ___.
9. Calculate the sum and record it in the math notebook.

   16 + 24 = 20
10. The full write-up in the math notebook is:

    4 x (4 + 6) =

    (4 x 4) + (4 x 6) =

    16 + 24 = 40
11. Select a prepared equation slip and continue as before.

## Aim

### Direct

Development of order, concentration, coordination, independence and exactness.

To memorize the multiplication facts.

To introduce algebraic formulations with multiplication.

## Age

7 to 8 years.

## Presentation 8 (Square of a Binomial)

1. Select a hundred square and name what it is: 10 x 10 = 100.
2. Record this equation in the math notebook.
3. Divide the square across the top into two parts, for example, 7 + 3.
4. Count across seven and place a rubber band vertically down the hundred square.
5. Count down seven and place a rubber band horizontally across the hundred square.
6. Indicate the seven square formed on the hundred square. Take the real square of seven and place it over the "mock" square of seven.
7. Indicate the three square formed on the hundred square. Take the real square of three and place it over the "mock" square of three.
8. Indicate the other two shapes that are formed: 3 x 7 on the right side of the square, and 7 x 3 on the bottom side of the square.
9. Place the real squares aside.
10. With the bead bars, reproduce the square of seven and the square of three, and place them in the same position as in the real square.

11. Make the rectangles 3 x 7 and 7 x 3 and place them in their relative position next to the squares. A new square has been formed.

12. Record this equation in the math notebook:

$$(7 \times 7) + (3 \times 3) + (3 \times 7) + (7 \times 3) = 100$$

$$49 + 9 + 21 + 21 = 100$$

13. Explain that you wish to rewrite the equation using a special sign. To indicate the square of a number, a 2 high off the line is written.

$$7^2 + 3^2 + (3 \times 7) + (7 \times 3) = 10^2$$

$$49 + 9 + 21 + 21 = 10^2$$

The raised two is called an 'exponent'.

14. Explain that when the same factors are multiplied there is an easier way to record it. Instead of:

$$(3 \times 7) + (7 \times 3)$$

it may be recorded:

$$2(3 \times 7)$$

15. With this information we may record the equation:

$$7^2 + 3^2 + 2(7 \times 3) = 10^2$$

16. Continue with the possible binomials:

$$9^2 + 1^2 + 2(9 \times 1) = 10^2$$

$$8^2 + 2^2 + 2(8 \times 2) = 10^2$$

$$6^2 + 4^2 + 2(6 \times 4) = 10^2$$

$$5^2 + 5^2 + 5^2 + 5^2 = 10^2$$

**Direct Aim:** Development of order, concentration, coordination, independence and exactness.

To memorize the multiplication facts.

To introduce algebraic formulations with multiplication.

**Age**

7 to 8 years.

**Presentation 9 (Square of the Trinomial)**

1. Select a hundred square and, with rubber bands, make a binomial as in the past presentation.
2. Explain that the binomial just completed divides the hundred square into two parts along each side. This produces four areas within the square.
3. Now we wish to divide the hundred square into three parts along each side. This produces nine areas within the square.
4. Divide it along the sides: 5 + 3 + 2.
5. Create these same areas with the three squares and the bead bars.
6. Analyze the figure moving across the square from left to right:

$$(5 \times 5) + (5 \times 3) + (5 \times 2)$$
$$25 + 15 + 10 = 50$$
$$(3 \times 5) + (3 \times 3) + (3 \times 2)$$
$$15 + 9 + 6 = 30$$
$$(2 \times 5) + (2 \times 3) + (2 \times 2)$$
$$10 + 6 + 4 = 20$$

100 Total

**NOTE:**

At a later date and after much practice with the squares of the binomial and trinomial, the mathematical expression is introduced.

Lay out the bead squares and bead bars. They mathematically express this:

$(5 + 3 + 2) \times (5 + 3 + 2) =$

$3^2 + 2(3 \times 5) + 2(3 \times 2) + 5^2 + 2(5 \times 2) + 2^2 =$

Recall that in constructing the square of a trinomial, the mathematical expression assumed nine operations and all nine operations were expressed. The new mathematical expression only requires six operations.

Binomial: Two terms and four operations.

Trinomial: Three terms and nine operations.

Quatrenomial: Four terms and sixteen operations.

**Direct Aim:**

Development of order, concentration, coordination, independence and exactness.

To memorize the multiplication facts.

To introduce algebraic formulations with multiplication.

**Age:**

7 to 8 years.

**Presentation 10 (Passing from One Square to a Consecutive Square)**

1. Select a square: six.

2. Say, "Let's go from this square of six to a square of seven."

3. The square of 6 is 36. The square of 7 is 49. To go from 36 to 49, thirteen beads are required. There is not a bead bar of thirteen to wrap around the six square to create a seven square.

4. Select two 7-bead bars, and place them along the sides of the six

square, and show how they are too long.

5. Select two six bars and place them along the sides of the six square. With the addition of a single one-bead in the corner, the 7 square is complete.

6. Count the number of beads added to the six square: 13.

7. Verbalize the process: "To go from 6 squared to 7 squared, 2 six-bead bars are added, plus one.

8. Record this in the math notebook: $6^2$ to $7^2 = 6^2 + 6 + 6 + 1$

9. The child may then begin with one and pass successively from one square to the next and record it in the math notebook:

$1^2$ to $2^2 = 1^2 + 1 + 1 + 1 = 1^2 + 3 = 4$

$2^2$ to $3^2 = 2^2 + 2 + 2 + 1 = 2^2 + 5 = 9$

$3^2$ to $4^2 = 3^2 + 3 + 3 + 1 = 3^2 + 7 = 16$

$4^2$ to $5^2 = 4^2 + 4 + 4 + 1 = 4^2 + 9 = 25$

$5^2$ to $6^2 = 5^2 + 5 + 5 + 1 = 5^2 + 11 = 36$

$6^2$ to $7^2 = 6^2 + 6 + 6 + 1 = 6^2 + 13 = 49$

$7^2$ to $8^2 = 7^2 + 7 + 7 + 1 = 7^2 + 15 = 64$

$8^2$ to $9^2 = 8^2 + 8 + 8 + 1 = 8^2 + 17 = 81$

$9^2$ to $10^2 = 9^2 + 9 + 9 + 1 = 9^2 + 19 = 100$

**NOTE:**

In each passage from one square to the next, the amount added increases by two.

**Presentation 11 (Passage from One Square to a Non-Consecutive Square)**

1. Select a square: six.

2. Say, "Let's go from this square of six to a square of eight.

3. The square of 6 is 36. The square of 8 is 64. To go from 36 to 64, twenty-eight beads are required. There is not a bead bar of twenty-eight to wrap around the six square to create an eight square.

4. Select four six-bead bars and place two bead-bars along each of two adjacent sides of the square.

5. Where the bead bars intersect is a two square. Select the two square and place it in the space.

6. Count the number of beads added to the six square: 28.

7. Verbalize the process: "To go from 6 squared to 8 squared, 4 six-bead bars are added plus a two square.

8. Record this in the math notebook:

$$6^2 \text{ to } 8^2 = 6^2 + 2(6 \times 2) + 2^2 =$$

$$36 + 24 + 4 = 64$$

9. The child may select squares and make them into non-consecutive squares.

# CONSTRUCTION OF THE DECANOMIAL

**Material**

A box with ten compartments containing fifty-five bead bars in each compartment 1 - 10.

**Presentation 1 (Vertical Construction)**

1. Say, "Let's construct the multiplication tables with the bead bars in a special way."
2. Lay out one bar for each of the numbers 1 - 10 across the top of the work area.
3. Say, "Let;s construct the table of one with the one-beads down the left column."
4. Continue with the table of two with the two-bead bars down the second column.
5. Continue with the rest of the columns.
6. Upon completion of the bead bar, lay out a chart on which to duplicate the pattern.

**NOTE:**

Multiplicand remains the same and the multiplier changes.

**Aim**

**Direct**

Development of order, concentration, coordination, independence and exactness.

To point out the interrelationship of the multiplication facts.

**Age**

6 to 8 years.

**Presentation 2 (Horizontal Construction)**

1. Say, "Let's construct the multiplication tables with the bead bars as before."
2. Lay out one bead bar for each of the numbers 1 - 10 across the top of the work area.
3. Then lay out the second row with two bead-bars.
4. Continue with the rest of the rows.
5. Upon completion of the bead bar layout, a chart may be used to duplicate the pattern.

**Presentation 3 (Angular Construction)**

1. Say, "Let's construct the multiplication tables with the bead bars in a special way."
2. Lay out one bar for each of the numerals 1 - 10 across the top of the work area.
3. Indicate the first bead and say, "This is 1 x 1." Indicate the second bead bar. "This is 1 x 2."
4. Indicating the space under the one-bead say, "Let's put 2 x 1 here," and place 2 ones.
5. Say, "Let's exchange the 2 ones for a two bar."
6. Indicate the position for 2 x 2 and say, "Let's put 2 x 2 here," and place 2 twos.
7. Indicate the third bead bar in the top row and say, "This is 1 x 3."
8. Indicate the space under the 2 x 1 and say, "Let's put 3 x 1 here," and place 3 ones.
9. Say, "Let's exchange the 3 ones for a three bar."
10. Indicate the space under 1 x 3 and say, "Let's put 2 x 3 here," and place 2 threes.
11. Indicate the space under the 2 x 2 and say, "Let's put 3 x 2 here," and place 3 two-bead bars.

12. Say, "Let's exchange the 3 two-bead bars for 2 three-bead bars."
13. Indicate the space for 3 x 3 and place 3 three-bead bars.
14. Continue with the table in this way.
15. Upon completion of 1 - 10, ask what geometric shape has been formed: the square.
16. Further analyze the geometric forms in the table:
    A. Two rectangles equal to each other along the two sides.
    B. Note the square of each number along the diagonal.
17. Exchange the "mock" squares for real squares.
18. Name a table and find it in the layout.
19. Have equation labels to place on rectangle or square groupings:

    2x 7 = ___.

20. Have product labels to place on rectangle or square groupings: 14.
21. Verbalize the equations on each line.

**Direct Aim**

The interrelationship of the multiplication facts.

**Age**

6 to 8 years.

**Presentation 4 (Numeral Decanomial)**

**Material**

Ten envelopes which contain cards of the tables from one to ten. The cards are arranged as follows: In the envelope marked 0, the perfect squares of the numbers one to ten on dark blue cards.

18 rectangles, 9 vertical and 9 horizontal, which correspond to the table of one.

16 rectangles, 8 vertical and 8 horizontal, which correspond to the table of two.

14 rectangles, 7 vertical and 7 horizontal, with one side equal to three.

12 rectangles, 6 vertical and 6 horizontal, one side equals 4.

10 rectangles - one side equals 5.

8 rectangles - one side equals 6.

6 rectangles - one side equals 7.

4 rectangles - one side equals 8.

2 rectangles - one side equals 9.

The products are written on each card. For example, on the card with the dimensions 1 x 2, the product 2 is written. On the cards for the squares, the number and the exponent are written in the upper left and the product is written in the middle of the card.

**Presentation**

1. Take the squares and lay them along a diagonal.
2. Take the table of ones and lay them in order. The horizontal cards are placed along the top row, the vertical are placed along the first column.
3. Continue with each table in this manner.
4. Variations of this layout:
    - A. Look for two rectangles of the same product and lay them out.
    - B. Any table may be selected and placed into position.
    - C. Mix up the rectangles and choose one at random to place in the appropriate position on the table.

**NOTE:**

Upon completion of the table of five, only 10 new multiplication equations remain. The construction of the decanomial is the construction of the Pythagorean Table.

The decanomial is easier than the square of the binomial and of the trinomial. Therefore, it may be presented first.

**Aim**

**Direct**

Development of order, concentration, coordination, independence and exactness.

The interrelationship of the multiplication facts.

**Age**

6 to 8 years.

**Presentation 5 (Substitution Game 1)**

1. Lay out the decanomial as in Presentation #3, in angular form.
2. Have all the squares for the numbers.
3. Replace all the "mock" squares with real squares.
4. Remove the one set of two above the two square and one set of two to the left of the two square. Replace these two bead-bars with a two square.
5. Continue in this manner 3 - 10, forming the squares.
6. Upon completion, the Pythagorean Table has been transformed into a square of squares.
7. Stack all the squares for each number, forming a cube for each. The cubes are placed along the diagonal as were the squares.
8. The "mock cubes" are exchanged for cubes of the numbers.
9. The cubes may be stacked into a tower.
10. The reverse process may be carried out: The cubes exchanged for squares, the squares exchanged for bead bars.
11. These groupings may be mathematically recorded.

**Aim**

**Direct**

The interrelationship of the multiplication facts.

**Age**

6 to 8 years.

# CONSTRUCTION OF THE DECANOMIAL, Continued

| | | | | | | | | | |
|---|---|---|---|---|---|---|---|---|---|
| 1 | x | 1 | = 1 | $1^2$ | = | 1 | $1^3$ | = | 1 |
| | | | | | | | | | |
| 2 | x | 2 | = 4 | $2^2$ | = | 4 | $2^3$ | = | 8 |
| 2 | x | 2 | = 4 | $2^2$ | = | 4 | | | |
| | | | | | | | | | |
| 3 | x | 3 | = 9 | $3^2$ | = | 9 | $3^3$ | = | 27 |
| 3 | x | 3 | = 9 | $3^2$ | = | 9 | | | |
| 3 | x | 3 | = 9 | $3^2$ | = | 9 | | | |
| | | | | | | | | | |
| 4 | x | 4 | = 16 | $4^2$ | = | 16 | $4^3$ | = | 64 |
| 4 | x | 4 | = 16 | $4^2$ | = | 16 | | | |
| 4 | x | 4 | = 16 | $4^2$ | = | 16 | | | |
| 4 | x | 4 | = 16 | $4^2$ | = | 16 | | | |
| | | | | | | | | | |
| 5 | x | 5 | = 25 | $5^2$ | = | 25 | $5^3$ | = | 125 |
| 5 | x | 5 | = 25 | $5^2$ | = | 25 | | | |
| 5 | x | 5 | = 25 | $5^2$ | = | 25 | | | |
| 5 | x | 5 | = 25 | $5^2$ | = | 25 | | | |
| 5 | x | 5 | = 25 | $5^2$ | = | 25 | | | |

| | | | | | | |
|---|---|---|---|---|---|---|
| 6 x 6 | = 36 | $6^2$ | = 36 | $6^3$ | = | 216 |
| 6 x 6 | = 36 | $6^2$ | = 36 | | | |
| 6 x 6 | = 36 | $6^2$ | = 36 | | | |
| 6 x 6 | = 36 | $6^2$ | = 36 | | | |
| 6 x 6 | = 36 | $6^2$ | = 36 | | | |
| 6 x 6 | = 36 | $6^2$ | = 36 | | | |
| | | | | | | |
| 7 x 7 | = 49 | $7^2$ | = 49 | $7^3$ | = | 343 |
| 7 x 7 | = 49 | $7^2$ | = 49 | | | |
| 7 x 7 | = 49 | $7^2$ | = 49 | | | |
| 7 x 7 | = 49 | $7^2$ | = 49 | | | |
| 7 x 7 | = 49 | $7^2$ | = 49 | | | |
| 7 x 7 | = 49 | $7^2$ | = 49 | | | |
| 7 x 7 | = 49 | $7^2$ | = 49 | | | |
| | | | | | | |
| 8 x 8 | = 64 | $8^2$ | = 64 | $8^3$ | = | 512 |
| 8 x 8 | = 64 | $8^2$ | = 64 | | | |
| 8 x 8 | = 64 | $8^2$ | = 64 | | | |
| 8 x 8 | = 64 | $8^2$ | = 64 | | | |
| 8 x 8 | = 64 | $8^2$ | = 64 | | | |
| 8 x 8 | = 64 | $8^2$ | = 64 | | | |
| 8 x 8 | = 64 | $8^2$ | = 64 | | | |
| 8 x 8 | = 64 | $8^2$ | = 64 | | | |

| | | | | | | | | | |
|---|---|---|---|---|---|---|---|---|---|
| 9 | x | 9 | = 81 | $9^2$ | = | 81 | $9^3$ | = | 729 |
| 9 | x | 9 | = 81 | $9^2$ | = | 81 | | | |
| 9 | x | 9 | = 81 | $9^2$ | = | 81 | | | |
| 9 | x | 9 | = 81 | $9^2$ | = | 81 | | | |
| 9 | x | 9 | = 81 | $9^2$ | = | 81 | | | |
| 9 | x | 9 | = 81 | $9^2$ | = | 81 | | | |
| 9 | x | 9 | = 81 | $9^2$ | = | 81 | | | |
| 9 | x | 9 | = 81 | $9^2$ | = | 81 | | | |
| 9 | x | 9 | = 81 | $9^2$ | = | 81 | | | |

| | | | | | | | | | |
|---|---|---|---|---|---|---|---|---|---|
| 10 | x | 10 | = 100 | $10^2$ | = | 100 | $10^3$ | = | 1000 |
| 10 | x | 10 | = 100 | $10^2$ | = | 100 | | | |
| 10 | x | 10 | = 100 | $10^2$ | = | 100 | | | |
| 10 | x | 10 | = 100 | $10^2$ | = | 100 | | | |
| 10 | x | 10 | = 100 | $10^2$ | = | 100 | | | |
| 10 | x | 10 | = 100 | $10^2$ | = | 100 | | | |
| 10 | x | 10 | = 100 | $10^2$ | = | 100 | | | |
| 10 | x | 10 | = 100 | $10^2$ | = | 100 | | | |
| 10 | x | 10 | = 100 | $10^2$ | = | 100 | | | |
| 10 | x | 10 | = 100 | $10^2$ | = | 100 | | | |

**NOTE:**

The square is formed by the first ten numbers:

$$1 + 2 + 3 + 4 + 5 + 6 + 7 + 8 + 9 + 10.$$

The following equation is sensorially verified:

$$(1 + 2 + 3 + 4 + 5 + 6 + 7 + 8 + 9 + 10)^2 =$$

$$1^3 + 2^3 + 3^3 + 4^3 + 5^3 + 6^3 + 7^3 + 8^3 + 9^3 = 10^3$$

## Aim

### Direct

Development of order, concentration, coordination, independence and exactness.

The interrelationship of the multiplication facts.

## Age

6 to 8 years.

# POWERS OF NUMBERS

**Material:**

The bead cabinet with the short and long chains, the squares and cubes, and one bead bar for each number, one through ten.

Two boxes labeled: Chain of the Square

Chain of the Cube

Each box has an envelope for each number, one through ten.

The arrows for each number follow this pattern:

For the chain of two, there are three arrows:

The first is labeled: 2

The second: 4, with 2 x 2 on the reverse side

The third: 4, with $2^2$ on the reverse side

For the chain of three, there are three arrows:

The first is labeled: 3

The second: 9, with 3 x 3 on the reverse side

The third: 9, with $3^2$ on the reverse side

The arrows for the chains of the cube follow this pattern:

For the chain of two, there are four arrows:

The first is labeled: 2

The second: 8, with 2 x 2 x 2 on the reverse side

The third: 8, with $2^2$ x 2 on the reverse side

The fourth: 8, with $2^3$ on the reverse side

For the chain of three, there are four arrows:

| | |
|---|---|
| The first is labeled: | 3 |
| The second: | 27, with 3 x 3 x 3 on the reverse side |
| The third: | 27, with $3^2$ x 3 on the reverse side |
| The fourth: | 27, with $3^3$ on the reverse side |

**NOTE:**

The arrow with the bsae is always smaller than the others.

**Presentation:**

1. Proceed as in the squaring chain and cubing chain presentation: a bead, a bar, the chain of the square, a square, the chain of the cube, and a cube. This is review for the child.

2. Place the material aside. Lay out the chain of a square (5) and the arrows.

3. The three arrows are labeled with:

   A. The base: 5.

   B. The product of the square: 25; on the reverse side the multiplicand and multiplier: 5 x 5.

   C. The product of the square: 25; on the reverse side the exponent for the square: $5^2$.

4. Place the narrow arrow with the base after the first set of bead bars.

5. Place the remaining two arrows at the end of the chain. They both have the product. On the other side of the arrows is the equation 5 x 5 and $5^2$. We have another way to write 5 sets of 5.

6. Explain to the child that the equation may be expressed by 5 x 5 of the shorthand $5^2$.

7. Lay out the chain of a cube (5) and the arrows.

8. The four arrows are labeled with:

    A. The base: 5.

    B. The product of the cube: 125; on the reverse side the multiplicands and multiplier: 5 x 5 x 5.

    C. The product of the cube: 125; on the reverse side the exponent: $5^2$ x 5.

    D. The product of the cube: 125; on the reverse side the shorthand: $5^3$.

9. Place the narrow arrow with the base after the first set of bead bars.

10. Place the remaining three arrows at the end of the chain. On one side all three arrows have the same product. On the other side of the arrows is the equation 5 x 5 x 5, $5^2$ x 5, and $5^3$.

11. Explain to the child that the equation may be expressed by 5 x 5 x 5 or $5^2$ x 5, or the shorthand $5^3$.

12. Help the child to arrive at the realization that: When we say $10^1$, we add one zero to the number 1. When we say $10^2$, we add two zeros to the number 1. When we say $10^3$, we add three zeros to the number 1. It is the same with other notation. We add as many zeros as the small raised number after the 10 indicates. This number is called an exponent. Example: When we say $10^8$, we add 8 zeros to the number 1.

**Direct Aim:**

To help the child understand the laws which govern the powers of numbers, and to make him aware of the first powers; that is, the first, second, and third power of the first 10 numbers of the natural series of numeration.

**Age:**

After 8 years.

**NOTE:**

With this material, Dr. Montessori says we want to give the child the possibility of learning the different ways in which numbers can be grouped and the different hierarchies of numbers. This way of grouping numbers is a conventional system established by man. We must try to help the child understand the fact that these are artificial groups established by man, and that in this system there is a rule which governs numeration: obedience to the law of the group. Each group is formed by single units. For example, three is formed by three units. Each number represents one set, one group. For example, the group two is formed by two single units and it represents one set. In higher math, one is designated as a set also. It is called the group of numerosity.

# FURTHER EXPLANATION – POWERS OF NUMBERS

### The First Power

The first power includes the numbers one to nine and the one with zero. This is the lowest in value of all the hierarchies, but it is important because the numbers in this level give value to those in the other hierarchies. What gives importance to one? The zero - so every time we add a zero to the one, its value increases.

### The Second Power

To form a number to the second power, we must take that number times itself:

$4^2$ is formed by 4 x 4.

$4^1 = 4$

$4^2 = 4 \times 4$

### The Third Power

In order to obtain the third power of a number, we must take the second power times the base: $4^3$ is formed by $4^2$ x 4.

$4^3 = 4^2 \times 4$

From this brief consideration of the first three powers of numbers, we see that the governing rule is the base. It is the entity which determines the value of the powers. Therefore, it is very important.

We see, also, that all the first powers are formed of units; the second powers are always squares; the third powers are always cubes.

If we want to represent the powers in geometric form, we will have:

$10^0$

$10^1$

$10^2$

$10^3$

With $10^3$, we have come back to the point. With $10^4$, we will have a line of cubes, with $10^5$, we will have a surface of cubes, and with $10^6$, we will have a cube of cubes. A later presentation, the Hierarchy of Numbers, takes the power of numbers to millions.

# INTRODUCTION TO THE UNIT DIVISION BOARD

**Material:**

A board similar to the multiplication board, having a green row across the top of the board with the numerals 1 - 9 representing the divisor. The 81 holes on the board represent the maximum dividend of 81. The numbers 1 - 9 along the left column represent the quotient.

Nine green skittles to represent the divisor along the top of the board.

Eighty-one green unit beads to represent the dividend to be placed in the holes.

Division paper on which to record the equations.

In any system of numeration, the maximum dividend is $(b - 1)^2$, i.e., (base one)$^2$ or with base equals ten: $(10 - 1)^2 = 81$.

The maximum divisor is (b - 1): 10 - 1 = 9.

**Presentation (Research of Dividends)**

1. Place the board in the middle of the table and examine it with the child.
2. Indicate the holes along the top that will hold the skittles. The skittles represent the divisor.
3. Indicate the smaller holes on the board. The green beads are placed in the holes, and they represent the dividend.
4. Select a prepared equation slip: 12 ÷ 4 = ___.
5. Record the equation in the math notebook.
6. Count out the twelve green beads to represent the dividend.

  

7. Place the four skittles along the top of the unit division board, beginning at the left to represent the divisor.
8. Distribute the twelve beads evenly, left to right, to the four skittles. Give one bead to each skittle and then ask, "Does each skittle have the same amount?" Then proceed to distribute a second bead to each skittle. Continue in this manner until all the beads are equally shared or distributed to the skittles.
9. Ask, "Does each skittle have the same amount?"
10. "The answer is the share of one, the amount of beads that one skittle has."
11. Record the quotient in the math notebook.

**Aim**

**Direct**

Development of order, concentration, coordination, independence and exactness.

To memorize the division facts.

**Age**

6 to 7 years.

# UNIT DIVISION BOARD

**Material**

Complete set of materials for the unit division bead board

Four-column division recording paper

Division Control Chart 1

**Presentation 1**

1. Examine the recording paper. Point out the heading, "Division" and the headings for the four columns: "Dividend," "Divisor," "Quotient," and "Remainder." Note that there are nine squares down each column; this is because with some dividends we will be able to do nine equations. With some dividends there will be few possibilities of division.
2. Dividend of 81: Take the box of 81 green beads and explain that you wish to distribute all 81 beads. Record the 81 under the dividend column of the paper.
3. Take nine skittles and place them along the top of the division board and explain that you wish to equally distribute the 81 beads to the skittles. Record the 9 under the divisor column of the paper.
4. Proceed to distribute the 81 beads, moving from left to right across the board, placing one bead under each skittle, then another, etc., until all the beads have been distributed.
5. W hen you have completed the distribution, show that the box of beads is empty.
6. Indicate that nine is the quotient, because there are nine beads in each individual set on the board.
7. Write the 9 in the column marked quotient.
8. Note that since there were no beads left in the box when you completed the

distribution, you have no remainder in this example.

9. Because there is no remainder, this is a very important equation, so it must be underlined in red.
10. In using this board there are two rules:
    - A. The quotient cannot be larger than nine.
    - B. The remainder cannot be equal to or larger than the divisor.
11. Ask the child if you can distribute 81 beads among 8.
12. Remove the ninth skittle and the nine beads that are below it.
13. Place these beads in the box, as there are no places under the first eight skittles.
14. Note that you cannot distribute the 81 beads among 8 skittles, because the maximum quotient is nine. There is nothing to do with the extra beads, which are more in number than the divisor, 8.
15. Dividend of 80: Take another paper on which to work with the dividend 80.
16. Remove one bead entirely.
17. Return the ninth skittle to the board, and tell the child that now you want to divide 80 by 9.
18. Record the dividend 80 and the divisor 9.
19. Add the beads to the ninth row. (It is not necessary to remove all the beads and do the complete distribution.)
20. There will be only 8 beads for the last skittle, so it will be necessary to remove one bead from each of the first eight skittles; this way the division will be even.
21. Note that the quotient is eight; record the 8 in the column for quotient.
22. Note that there are eight beads which remain in the box, so this 8 is the remainder.
23. Record the remainder 8 in the column for the remainder.
24. Remove one skittle and the beads under it, as was done with the dividend 81.
25. 80 - 8 is not possible, because it results in a quotient of 10.

26. Dividend of 79: Remove one more bead and work with the dividend 79, as you have done with the other dividends, taking a new paper to record the work.

**NOTE:**

After this much direction, the child proceeds along with the other dividends to 1, recording the work. Explain that you are especially interested in those equations in which there is no remainder. The equations with no remainder should be underlined in red because these are the equations which will have to be memorized. When the child goes on to other materials for the memorization of division, those equations which were underlined in this preliminary work are the ones which will need to be memorized; the other equations will not appear. Of the 81 dividends, only 36 are memorized. This process is a great deal of work, and hardly any child wants to complete it entirely. It is not necessary to do all the dividends to one; it is necessary only to do enough to understand the procedure used to arrive at the printed booklets. After working with some dividends, show the set of pages prepared by the teacher.

**Aim**

**Direct**

Development of order, concentration, coordination, independence and exactness.

To memorize the division facts.

**Age**

7 to 8 years.

## DIVISION

| Dividend | Divisor | Quotient | Remainder |
|---|---|---|---|
| | | | |

## DIVISION

| Dividend | Divisor | Quotient | Remainder |
|---|---|---|---|
| | | | |

| Dividend | Divisor | Quotient | Remainder |
|---|---|---|---|
| | | | |

| Dividend | Divisor | Quotient | Remainder |
|---|---|---|---|
| | | | |

**Presentation 2 (Booklet of Prepared Equations)**

1. Examine the booklet, noting that there is a new page for each dividend and each dividend is divided by nine through one.
2. Turn to dividend 5.
3. Establish the dividend by counting out five beads.
4. Read the first equation: $5 \div 5 =$ ___. Place the five skittles representing the divisor at the top of the board.
5. Distribute the beads equally across the board. The quotient is one because the share of one skittle is one.
6. Record the quotient in the booklet: $5 \div 5 = 1$.
7. Read the next equation: $5 \div 4 =$ ___.
8. Remove one skittle from the board.
9. Distribute the five beads.
10. Each skittle receives one bead, and one bead remains which cannot be distributed.
11. Record the quotient and the remainder: $5 \div 4 = 1$ R 1.
12. Proceed with the rest of the equations on the 5 dividend page.
13. The equations are checked for accuracy with the Division Control Chart 1.
14. The numbers in the white squares represent the quotients resulting from division in which there is no remainder.
15. Division with a remainder may be calculated from the chart:
    A. Locate the nearest quotient to the right.
    B. Subtract.
    C. The difference is the remainder.
16. Point out to the child the numbers 7 - 5 - 3 - 2 - 1 on the line of dividends. These are the only dividends on white squares. Explain that these numbers are divisible only by one and themselves. Give the term "prime number," and

explain that these are the most important prime numbers, although there are other prime numbers the child will discover.

**Aim**

**Direct**

Development of order, concentration, coordination, independence and exactness.

To memorize the division facts.

**Age**

7 to 8 years.

# DIVISION BEAD BOARD

**Material**

Complete set of materials for the division bead board

Division prepared equation slips

Division Control Chart 1

**Presentation**

1. Select an equation: 21 ÷ 7 = ___.
2. Record the equation in the math notebook.
3. Establish the dividend by counting out 21 beads.
4. Establish the divisor by laying out 7 skittles along the top of the board.
5. Distribute the dividend.
6. Record the quotient, the share of one: 21 ÷ 7 = 3.
7. Clear the board, removing beads and skittles.
8. Choose another equation.
9. Proceed to do this equation as before.
10. Upon completion of the prepared equation slips, the equations are checked for accuracy by referring to Division Control Chart 1.

**Aim**

**Direct**

Development of order, concentration, coordination, independence and exactness.

To memorize the division facts.

**Age**

7 to 8 years.

# DIVISION CHART 1

**Material**

Prepared equation slips

Division Chart 1

Math notebook

**Presentation**

1. Examine Division Chart 1. The dividend is along the top; the divisor is along the left diagonal; the quotients appear in the matrix of the chart.
2. Select a prepared slip.
3. Record the equation in the math notebook.
4. With the right index finger, find the dividend at the top of the chart.
5. With the left index finger, find the divisor along the left diagonal of the chart.
6. Bring the right index finger down and the left index finger across. The square in which the fingers meet is the quotient.
7. Record the quotient in the math notebook.
8. Upon completion, the equations are checked for accuracy by referring to Division Control Chart 1.

**Aim**

**Direct**

Development of order, concentration, coordination, independence and exactness.

To memorize the division facts.

**Age**

7 to 8 years.

 

# DIVISION CHART 2

**Material**

Prepared equation slips

Quotient stamps

Division Control Chart 1

Division Chart 1 and 2

Math notebook

**Presentation 1**

1. Examine Division Chart 2. It is a blank chart with the matrix containing no quotients.
2. Lay all the stamps face up on the table.
3. Select an equation and record it in the math notebook.
4. From the stamps, select the correct quotient and place it at the top of the Division Chart 1.
5. With the index finger of the right hand, find the dividend along the top of the chart.
6. With the index finger of the left hand, find the divisor along the left hand side of the chart.
7. Move the index finger down and the left index finger across. Where they meet is the blank square where the guotient stamp should be placed.
8. Record the quotient in the math notebook.
9. Select another prepared slip and proceed as before.

10. Upon completion of Division Chart 2, compare it to Division Chart 1. They should be identical. Compare the equations in the math notebook to Division Control Chart 1 for accuracy.

**Aim**

**Direct**

Development of order, concentration, coordination, independence and exactness.

To memorize the division facts.

**Age**

7 to 8 years.

**Presentation 2**

1. Lay all the stamps face up on the table.
2. Select a quotient (stamp) at random and place it at the top of Division Chart 2. Record it in the math notebook.
3. The child makes up an equation to satisfy the quotient and enters the complete equation in the math notebook.
4. With the index finger of the right hand, find the dividend along the top of the chart.
5. With the index finger of the left hand, find the divisor along the left diagonal of the chart.
6. Move the right index finger down and the left index finger across. Where they meet is the blank square where the quotient stamp should be placed.
7. Select another quotient stamp and proceed as before.
8. Upon completion of Division Chart 2, compare it to Division Chart 1. They

should be identical. Compare the equations in the math notebook to Division Control Chart 1 for accuracy.

## Aim

### Direct

Development of order, concentration, coordination, independence and exactness.

To memorize the division facts.

## Age

7 to 8 years.

## Presentation 3

1. The child lays all the stamps face up in stacks of the same denomination, from low numerals to high numerals.
2. Select a quotient stack (stamps).
3. Make up all the equations that would satisfy this quotient. The number of stamps in the stack indicates the number of possible equations.
4. Enter each equation in the math notebook.
5. With the index fingers, locate the positions of the quotients on the Division Chart 2.
6. Upon completion of the stacks of quotients (stamps), compare the filled in Chart 2 with Chart 1. They should be identical. Compare the equations in the math notebook to Division Control Chart 1 for accuracy.

## Aim

### Direct

To memorize the division facts.

## Age

7 to 8 years.

# DIVISION STORY PROBLEMS

**Material**

Generate story problems which ask for one of each of the factors in a division equation:

Dividend:

Jeff brings Halloween treats to share. He shares his treats equally with 4 children. Each child receives 5 treats. How many treats did Jeff bring?

Divisor:

Sam brings 42 crayons to share equally with his friends. Each child receives 7 crayons. How many children did Sam share his crayons with?

Quotient:

Shirley has 8 pumpkins. She shares the pumpkins equally with 4 friends. How many pumpkins does each friend receive?

**Presentation**

1. Select a story problem and read it.
2. Ask:
   - A. How many are there before anything is shared? What is the dividend?
   - B. How many are shared with? What is the divisor?
   - C. How many does each get? What is the quotient?
3. As the questions above are asked, the following blanks may be filled in:

   ___ ÷ ___ = ___.
4. The unfilled blank is the factor to be determined in the equation.
5. Record the equation in the math notebook.
6. Upon completion of the story problems, the Division Control Chart 1 is used to compare to the answers in the math notebook for accuracy.

## Aim

### Direct

Development of order, concentration, coordination, independence and exactness.

To place the division facts into the context of practical life experience.

## Age

7 to 8 years.

# RELATIONSHIPS OF MULTIPLICATION AND DIVISION

**Material**

Multiplication Chart 4

**Presentation**

1. Cover the last row on the chart, which is the table of ten. There are now 45 products on the chart.
2. Cover any repeats of numbers, on the pink or white squares. The only numbers left are the dividends in the combination booklets.
3. This activity links division with multiplication. We are proceeding from product to dividend, because multiplication is always linked to division. It is because of this relationship between multiplication and division that we are able to go from product to dividend.

**Aim**

**Direct**

Development of order, concentration, coordination, independence and exactness.

To establish the relationship between division and multiplication.

**Age**

7 to 8 years.

# ORAL GAMES FOR THE MEMORIZATION OF DIVISION

**Material**

Division equation slips

Quotient stamps

**Presentation 1**

1. The child selects a division equation slip and reads it.
2. Another child says the quotient.
3. The first child determines the accuracy of the response from memory or from Control Chart 1.
4. Proceed with the other equation slips in this manner.

**Aim**

**Direct**

Development of order, concentration, coordination, independence and exactness.

To memorize the division facts.

**Age**

7 to 8 years.

**Presentation 2**

1. The child selects a quotient (stamp) and reads it.
2. Another child says the possible equations to satisfy this quotient.

3. The first child determines the accuracy of the response from memory or from Control Chart 1.
4. Proceed with the other quotient stamps in this manner.

**Aim**

**Direct**

Development of order, concentration, coordination, independence and exactness.

To memorize the division facts.

**Age**

7 to 8 years.

# BASIC FORMATS FOR DIVISION

**Material**

Prepared slips with all possible equations for each of the following formats:

| | | | | | |
|---|---|---|---|---|---|
| ___ | ÷ | 7 | = | 3 | (Calculate for the dividend.) |
| 3 | = | ___ | ÷ | 7 | (Calculate for the dividend.) |
| 21 | ÷ | ___ | = | 3 | (Calculate for the divisor.) |
| 3 | = | 21 | ÷ | ___ | (Calculate for the divisor.) |
| ___ | = | 21 | ÷ | 7 | (Calculate for the quotient.) |
| 3 | = | ___ | ÷ | ___ | (Calculate for the dividend and divisor.) |

**Presentation**

1. Select a prepared equation slip of the first form at and record it in the math notebook.
2. Determine the factor that is not given and record it in red.
3. Upon co mpletion of the prepared equation slips, the Division Control Chart 1 is used to compare to the answers in the math notebook for accuracy.
4. Continue in this manner for each of the equation form ats.
5. The child may generate his own equations.